DISSOCIATION IN
HEAVY PARTICLE COLLISIONS

WILEY-INTERSCIENCE SERIES IN ATOMIC AND MOLECULAR COLLISIONAL PROCESSES

Advisory Editor

C. F. BARNETT

ION-MOLECULE REACTIONS *by E. W. McDaniel, V. Čermák, A. Dalgarno, E. E. Ferguson, L. Friedman*

THEORY OF CHARGE EXCHANGE, *by Robert A. Mapleton*

DISSOCIATION IN HEAVY PARTICLE COLLISIONS *by G. W. McClure, J. M. Peek*

EXCITATION BY HEAVY PARTICLE COLLISIONS *by E. W. Thomas*

In preparation:

Ionization by Heavy Particle Collisions by J. W. Hooper, F. W. Bingham
Charge Exchange in Gases and Solids by C. F. Barnett, I. A. Sellin

DISSOCIATION IN HEAVY PARTICLE COLLISIONS

G. W. McClure and J. M. Peek

Sandia Laboratories
Albuquerque, New Mexico

WILEY-INTERSCIENCE

A Division of John Wiley & Sons, Inc.
New York • London • Sydney • Toronto

Library of Congress Cataloging in Publication Data

McClure, G. W. 1923–
    Dissociation in heavy particle collisions.

    (Wiley-Interscience series in atomic and molecular collisional processes)
    Includes bibliographical references.
    1. Collisions (Nuclear physics) 2. Dissociation.
I.   Peek, James Mack, 1933–   joint author.
II. Title.

QC794.M175   539.7'54   73-37435
ISBN 0-471-58165-8

Printed in the United States of America.

10 9 8 7 6 5 4 3 2 1

Annex
539.754

SERIES PREFACE

The proliferation of scientific literature in the last two decades has exceeded the point where an individual scientist can either absorb or keep up with all of the publications even in narrow fields. In recognition of this situation, the Federal Council for Science and Technology, in 1963, established the National Standard Reference Data System with responsibility for its implementation assumed by the National Bureau of Standards of the Department of Commerce. In 1968, the existing authorities of the Department of Commerce to operate the NSRDS were supplemented by special legislation passed by the Congress of the United States. Within NBS, the standard reference data program is administered by the Office of Standard Reference Data which provides coordination and planning for data compilation efforts in the field of the physical sciences. As part of this effort, the Office of Standard Reference Data and the Division of Research of the United States Atomic Energy Commission have sponsored jointly the establishment of the Atomic and Molecular Processes Information Center at the Oak Ridge National Laboratory. The objectives of this Center are the collection, evaluation, and critical review of data in specialized topics of heavy particle atomic and molecular collisions. This monograph, *Dissociation in Heavy Particle Collisions*, is the third of a series of critical reviews. Grateful acknowledgment is made to Dr. G. W. McClure and Dr. J. M. Peek for their effort in making this monograph possible.

C. F. Barnett

PREFACE

This book is intended to aid scientists, engineers, and students who require information on the dissociation of molecules induced by collision with other "heavy" particles, including atoms, molecules, and ions. It is an index to the scientific literature and a guide to the understanding and evaluation of the literature rather than a data compilation.

The material falls into three main parts. Chapter 1 provides a general introduction to the subject of dissociative collisions. Chapters 2 through 6 discuss the experimental approach used to obtain dissociation cross section data, and Chapter 7 discusses the theoretical approach employed to obtain dissociation data. The experimental and theoretical portions of the text are self-contained and may be read essentially independently of one another according to the reader's specific interests.

The purposes of the experimental portion of the book are to provide a description of experimental methods, a complete index to the experimental data, a comprehensive discussion of the factors that need to be considered in evaluating a measurement, and a summary of the main qualitative results of the experimental work. It is hoped that the discussions in Chapters 5 and 6 give substantial aid to workers who are planning dissociation measurements. These chapters provide general comments on experimental errors and a detailed listing of the factors to be considered in planning a well-defined measurement with a meaningful error analysis.

We have restricted the scope of the literature search on experimental studies to cases in which the collision energy was well-defined. Almost all of the experimental data which meet this criterion have been obtained by the beam-gas method. This discussion of criteria for measurements applies mainly to this technique, but many of the developed criteria also apply to the crossed-beam method, which will no doubt come into increasing use in the near future. The reader may wish to skip from Chapter 3 to Chapter 5 on a first reading in order to maintain the continuity of the discussion on methods of measurement and evaluation criteria.

Chapter 7 is devoted to the theoretical studies of dissociation collisions between heavy particles. The construction of this chapter is similar to that used to discuss the experimental literature. The notation and theoretical models are introduced. The classical binary and first Born theories are discussed in some detail since they are used in around 85% of the theoretical papers containing cross section data. The reader is referred to the extensive literature concerned with the treatment of inelastic collisions in simpler systems for an understanding of the errors inherent in these approximate theories; the situation should not be markedly different for dissociation collisions. However, there are a number of approximations peculiar to the treatment of dissociative collisions, and the discussion of these subsidiary approximations, including a qualitative error analysis, constitutes a major portion of Chapter 7. Tables supply the reader with a list of papers concerned with a given collisional system, an identification of the theoretical framework used in each paper, and a list of approximations made in addition to the assumed theoretical framework. The discussion and tables should provide a critical perspective of both the achievements and needs in the theoretical study of dissociative collisions.

It is a unique feature of this book that almost every author of an experimental or theoretical paper on dissociation has been contacted for recent thoughts on the validity of the data he has published. Comments that have been received from the authors appear in the Appendix.

We wish to thank C. F. Barnett of the Atomic and Molecular Processes Information Center for suggesting this review and for supplying valuable advice and assistance throughout the course of the work. Many discussions with C. F. Barnett, J. W. Hooper, E. W. Thomas, and F. W. Bingham were helpful in organizing the presentation and establishing the evaluation criteria for measurements. A major portion of the section on thermal effects was written by F. W. Bingham and is used with Dr. Bingham's permission. The work on this volume could not have been completed without extensive help on the references and manuscript from Mrs. Vikki Tuttle, Mrs. Emma Dale Daniel, Mrs. Janice Kowalski and Mrs. Donna Cobble. Constructive comments by T. A. Green and A. Russek on the review of theoretical investigations and by H. B. Gilbody and G. H. Dunn on the review of experimental investigations are gratefully acknowledged. The support and encouragement of R. S. Claassen, J. R. Banister, and C. R. Mehl are greatly appreciated.

G. W. McClure
J. M. Peek

Albuquerque, New Mexico
October 1971

CONTENTS

Chapter 6
Qualitative Conclusions Based on Experimental Data   121

Chapter 7
Theory of Dissociative Collisions Between Heavy Particles   132

DISSOCIATION IN
HEAVY PARTICLE COLLISIONS

CHAPTER 1

INTRODUCTION

Information on dissociation of molecules under atom and molecule impact is required in a variety of technological and scientific applications. Among these are the assessment of radiation effects on molecular gases, the study of the behavior of molecular gases under high-voltage breakdown conditions, the deposition of energy in molecular gases by atomic and molecular beams, the production and destruction of beams in accelerators, and the injection of ions into fusion reactors.

The many applications for dissociation data have stimulated extensive experimental study of dissociative collisions during the past two decades. During that period about 200 individual journal articles, reports, and doctoral theses have been written on the subject. This intensive development has paralleled the acquisition of a vast quantity of data on the related topics of ionization, excitation, and charge exchange in collisions of atoms and molecules.

The species for which data exist range from H_2^+, the simplest molecule, to species comprising 14 atoms such as C_4H_{10} and B_5H_9. The quantity of data available on H_2^+ dissociation decisively dominates the overall body of information on dissociation, but a large collection exists for many heavier species.

Earlier reviews of work on atomic collisions, such as those of Massey and Burhop,[1] Hasted,[2] and McDaniel,[3] touch only lightly on the subject of collisional dissociation. None of these works reviews or lists more than a few percent of the total data now available and none places emphasis on the unusual array of experimental difficulties that dissociation measurements entail. The review of experimental techniques presented by Barnett and Gilbody[4] notes many of the important problems associated with the beam-gas collision technique which is the most widely used method of conducting dissociation measurements.

It is the purpose of this review to describe the experimental techniques employed in dissociation studies, to provide a convenient index to the

1

entire body of experimental data, to list and explain criteria to be observed in evaluating and planning dissociation measurements, to present
a selection of qualitative conclusions based on the experimental data,
and to review the theoretical understanding of dissociative collisions.

A major purpose of this book is to explore the experimental problems
in detail and to list a number of criteria for the conduct of well-defined
measurements coupled with well-defined error analyses. In writing the
material on this topic, we have attempted to summarize the cumulative
experience that we have gained in studying all of the papers containing
experimental data.

We have set a lower limit of 10 eV on the collision energy for the
purposes of this review. Very few studies yield useful data below this
energy. There are some data from swarm-type experiments that we have
excluded by this energy cutoff; however, in swarm experiments the collision energies are not well defined and even the identification of reactants
is sometimes in question.

It was our initial intention to present almost all of the data from
original sources, but difficulties were experienced in attempting to write
proper captions for data removed from their context. In nearly every
case the definition of the measured quantities was impossible to convey
clearly in a wordage substantially less than that of the full text of
the paper. This need for lengthy captions was due in some instances
to insufficient care on the part of the investigators to state succinctly
what was measured, and at other times to insufficient care in choosing
physical aspects of the apparatus so that a readily defined result was
achieved. In a few situations authors provided collimator slit dimensions
and other pertinent geometrical details but did not calculate the effective
or limiting solid angles for the benefit of the reader. We decided not
to undertake this task. Instead, we have attempted to spell out the
mathematical basis for deriving average, total, and differential cross
sections from the directly measured quantities in order that interested
nonspecialist readers may know how to formulate questions if they wish
to delve further by direct author contact.

Chapter 2 of this review presents a general discussion of the phenomenology of heavy-particle collisions including a mathematical definition of a measured cross section. Chapter 3 describes the experimental
methods most widely used in dissociation cross section measurements.
Chapter 4 provides a list of the molecular ion and molecular neutral
species for which experimental numerical data have been found, followed
by a detailed index to the literature according to the subject of the investigation. Chapter 5 presents the criteria to be observed in the conduct
of well-defined measurements. Chapter 6 discusses our conclusions based

on the experimental data. Chapter 7 reviews the theoretical work on collision-induced dissociation.

In conducting our search of the experimental literature, we have attempted to convey a complete picture of the information available in published form prior to January 1, 1970. Almost every author of data-bearing papers in our bibliography has been contacted by letter to make sure that our listings are complete and to solicit information regarding definition or accuracy not provided in the original paper. Appendix I lists supplemental information obtained through this correspondence.

Before referring to original papers for data and other detailed information, readers may find it helpful to refer to Appendix I and to the author index in which locations of all comments concerning individual papers in this review are listed.

About 20% of the authors did not respond to our inquiries. For this reason, and because some authors were contacted in 1967 and 1968 when an earlier deadline was set, we lack direct information from some authors for the completeness of our listings of their work on dissociation as of the January 1, 1970, deadline. We have tried to compensate to some degree by a careful check of all citations in the collected papers on dissociation to ascertain that we had complete coverage of literature containing experimental data. Cited references in all of the data-bearing papers were checked for possible content of information on techniques as well as numerical data. Many of the references listed in the bibliography relate to technique only.

CHAPTER 2

HEAVY PARTICLE COLLISIONS

2.1 PHENOMENA

In this section we consider the types of processes that give rise to structural modifications when two atoms or molecules (heavy particles) collide, and we enumerate the various products that may ensue. A thorough evaluation of a dissociation measurement requires a review of all of the possible interactions and a consideration of the possible interfering effect of each on the data under scrutiny.

We wish to consider binary encounters between pairs of particles, called reactants, of the following kinds:

1. Atomic ions.
2. Molecular ions.
3. Atoms.
4. Molecules.

In such collisions both reactants generally change their speed and direction of motion. In addition, one or more of the following processes may occur to alter the structure of the reactants:

1. Ionization.
2. Excitation.
3. Dissociation.
4. Electron transfer from one reactant to the other.

The probabilities of these processes are generally dependent on the particular reactant pair and the relative velocity of the reactants.

The following reactions of a singly charged methane ion with an argon atom or hydrogen molecule illustrate these processes:

$$CH_4^+ + Ar \rightarrow CH_4^+ + Ar^+ + e$$
$$CH_4^+(\nu) + Ar \rightarrow CH_4^+(\nu') + Ar$$

$$CH_4^+(\nu) + Ar \rightarrow CH_4^+(\nu') + Ar^*$$
$$CH_4^+ + Ar \rightarrow CH_3^+ + H + Ar$$
$$CH_4^+ + H_2 \rightarrow CH_4^+ + H + H^+ + e$$
$$CH_4^+ + H_2 \rightarrow CH_4^+ + H^+ + H^+ + 2e$$
$$CH_4^+ + A \rightarrow CH_4 + A^+$$
$$CH_4^+ + A \rightarrow CH_3 + H + A^+$$

The first reaction scheme represents ionization of the argon atom by removal of one electron. The second and third reactions represent a methane ion undergoing a transition from a vibrational state ν to a state ν', or an argon atom in the ground state undergoing a transition to an excited electronic state Ar*. The fourth reaction illustrates the dissociation of a CH_4^+ ion into a CH_3^+ ion and an H atom. The fifth and sixth schemes demonstrate two modes of collision-induced dissociation of an H_2 molecule resulting from impact with a CH_4^+ ion. The last two reaction schemes illustrate electron transfer from an argon atom to a methane ion.

A complete description of the reactants requires the specification of all of the following:

1. Atomic composition of chemical elements of each reactant species.
2. Isotopic composition of chemical elements of each species.
3. Molecular structure designation if isomeric states of reactant molecules exist.
4. Population distribution of reactants among various possible internal degrees of freedom, that is, electronic, vibrational, and rotational modes.
5. Orientation of magnetic moment relative to beam direction.

Given a specific reactant pair and a specific relative velocity, several of the processes listed above may occur either individually or simultaneously. Some may be forbidden in principle by the structure of the reactants or by the application of energy or momentum conservation laws. For example, dissociation is obviously possible only if one of the two reactants is a molecule; none of the processes can occur if both reactants are completely stripped atoms; scattering at certain angles may be forbidden for one of the reactants by the requirements of energy and momentum conservation; and excitation to a particular quantum state may be forbidden because the total energy of the reactants in the center of mass system before the collision is insufficient to allow the state in question to exist as a product state. Except for products that are forbidden by these and other more subtle theoretical considerations, one may expect any or all of the following types of collision products to emerge from a region where the reactants collide:

1. Atomic ions.
2. Molecular ions.
3. Neutral atoms.
4. Neutral molecules.
5. Electrons.
6. Photons.

The first four products, except bare nuclei in category 1, may have a variety of internal quantum states, kinetic energies, and directions of motion. Similarly, the emitted electrons may have a variety of kinetic energies and directions of motion. The energy spectrum of electrons emitted in a given direction may consist of "lines" associated with Auger processes, as well as a "continuous" part. Photons may, in general, be emitted in all directions and possess both line spectra and continuous spectra.

A structural modification of one reactant can be statistically correlated with a structural modification of the other reactant. For example, the probability that a molecular reactant will undergo dissociation may depend strongly on the specification of the final state of the other reactant. It is possible in principle to express the production probability of a given pair of final states of both particles as a cross section (see Sections 2.2–2.5); however, experiments have rarely dealt with the detailed final state specification of both particles.

In the present state of the experimental arts for the study of heavy particle collisions, the following kinds of measurements are common and have been conducted in many laboratories:

1. The total yield of detached electrons and/or negative ions.
2. The velocity distribution of detached electrons emitted in a given direction.
3. The total yield of slow charged particles—ions and electrons—resulting from changes in charge of either of the two reactants, including charged dissociation fragments.
4. The yield of ions of a particular mass-to-charge ratio and restricted velocity range in a given direction relative to the reactant velocities. Such ions may include scattered reactants—with or without changes in charge—and charged dissociation fragments.
5. The velocity distribution of a selected species of ion.
6. The yields of neutral-atom and neutral-molecule products in special cases.

7. The yield of photons belonging to a particular discrete transition in one of the atomic or molecular products.

8. The energy spectrum of photons emitted in a given direction.

9. The probability for simultaneous scattering of two reactants in two particular directions—sometimes called the coincidence method.

10. The determination of the relative frequency of various modes of fragmentation in dissociation of simple molecules.

Excellent introductions to many of these topics can be found in the reviews of Massey and Burhop,[1] Hasted,[2] McDaniel,[3] and Bederson and Fite.[4]

2.2   CROSS SECTION MEASUREMENTS

Probabilities of various kinds of collision process are usually expressed in terms of collision cross sections. Cross sections are of two broad types: total and differential. A total cross section usually refers to the total probability of producing a certain species of collision product, whereas a differential cross section usually refers to the production of a certain species of collision product with definite restrictions on one or more properties of the product describable in terms of continuous variables. In the discussion to follow we denote such properties by the set of variables $\mathbf{s} = \{s_1, s_2 \ldots s_n\}$ where the members s_i may represent emission angles, velocities, energies, wavelengths, and so forth. The number of members in the set may be one or many depending on the number of aspects of the collision event that the differential cross section is to describe.

It is helpful to represent a collision event as a point in the n-dimensional space of collision events with coordinates equal to the values of the n variables composing $\mathbf{s}$. A true differential cross section would refer to events in an infinitesimal volume of $\mathbf{s}$ space. Such a cross section is never actually measured. Instead, one measures events occurring in a finite volume of $\mathbf{s}$ space and normalizes the result to unit volume in $\mathbf{s}$ space to obtain an approximation to the differential cross section at a certain point $\mathbf{s}$. Similarly, a true total cross section is rarely measured. Actual measurements almost always exclude some portion of $\mathbf{s}$ space such as that represented by a group of products having a velocity or an emission direction not acceptable by the detector. In either type of measurement it is important to establish the region of $\mathbf{s}$ space involved. These ideas are expanded in the following discussion.

Let two interacting species be denoted a and b and suppose all type a particles have velocities $\mathbf{v}_a$ and all type b particles have velocity $\mathbf{v}_b$. Consider the relation expressing the interactions in an element of volume where both types of particle are present and are colliding with each other to produce product particles of type c having properties $\mathbf{s}$. The differential detector signal is related to the cross section and apparatus conditions as follows:

$$dI_c = N_a(\mathbf{x})N_b(\mathbf{x})\, d\mathbf{x}\, v \int_\tau \epsilon(\mathbf{x},\, \mathbf{a},\, \mathbf{s})\sigma(\mathbf{s})\, d\mathbf{s} \qquad (2.1)$$

where dI_c = the number of product particles c received by the detector per unit time from volume $d\mathbf{x}$.

$\quad \mathbf{x}$ = spatial coordinates $\{x,\, y,\, z\}$ of a particular point in the interaction region.

$\quad d\mathbf{x}$ = a volume element $dx\, dy\, dz$ in the interacting region.

$\quad N_a(\mathbf{x})$ = the density of type a particles at $\mathbf{x}$.

$\quad N_b(\mathbf{x})$ = the density of type b particles at $\mathbf{x}$.

$\quad v$ = the magnitude of the difference $\mathbf{v}_a - \mathbf{v}_b$.

$\quad \mathbf{s}$ = a set of properties $\{s_1,\, s_2,\, \ldots\, s_n\}$ of the product particle with respect to which the measurement is discriminatory.

$\quad d\mathbf{s}$ = a volume element $ds_1\, ds_2\, \ldots\, ds_n$ in the n-dimensional space of $\{s_1,\, \ldots\, s_n\}$.

$\quad \sigma(\mathbf{s})$ = the differential cross section corresponding to properties $\mathbf{s}$.

$\quad \mathbf{a}$ = a set of variable experimental parameters $\{a_1\, \ldots\, a_k\}$ that affects the portion of $\mathbf{s}$ space to which the detector responds.

$\quad \epsilon(\mathbf{x},\, \mathbf{a},\, \mathbf{s})$ = the efficiency of the detector for receiving products characterized by $\mathbf{s}$ originating at point $\mathbf{x}$ when the apparatus is set in configuration $\mathbf{a}$.

$\quad \tau$ = the region of $\mathbf{s}$-space in which $\epsilon \neq 0$.

In all practical situations the total signal received by the product particle detector originates in an interaction region of finite size. Values of N_a, N_b, and ϵ are usually functions of position $\mathbf{x}$ within that region. Hence the total detector signal is the integral with respect to $d\mathbf{x}$ of Eq. 2.1:

$$I_c = v \iint_{V\tau} N_a(\mathbf{x})N_b(\mathbf{x})\epsilon(\mathbf{x},\, \mathbf{a},\, \mathbf{s})\sigma(\mathbf{s})\, d\mathbf{s}\, d\mathbf{x} \qquad (2.2)$$

where V denotes the volume of $\mathbf{x}$ space characterized by points at which $N_aN_b \neq 0$. We have assumed that v is a constant independent of $\mathbf{x}$ and have placed it outside the integral. For some points in V, ϵ may be zero. Those regions do not contribute to the integral. It is generally true that $\epsilon(\mathbf{x},\, \mathbf{a},\, \mathbf{s})$ for fixed $\mathbf{x}$ and $\mathbf{a}$ has a value other than zero over a finite volume of s space, so that the signal is not characterized by a point in s space but a finite volume of s space. The function of $\mathbf{a}$ and $\mathbf{s}$ given by

$$F(\mathbf{a}, \mathbf{s}) \equiv \int_V N_a N_b \epsilon \, d\mathbf{x} \tag{2.3}$$

defines the resolution of the measurement; it establishes the weighting given by the measurement to every point of s space for any $\mathbf{a}$. The measured signal I_c represents the average value of $\sigma(\mathbf{s})$ with the function $F(\mathbf{a}, \mathbf{s})$ as weighting factor. The average value of the differential cross section in the region of s space accessible in configuration $\mathbf{a}$ is defined as

$$\bar{\sigma}(\mathbf{a}) \equiv \frac{\displaystyle\iint_{V_\tau} N_a N_b \epsilon \sigma \, d\mathbf{s} \, d\mathbf{x}}{\displaystyle\iint_{V_\tau} N_a N_b \epsilon \, d\mathbf{s} \, d\mathbf{x}} \tag{2.4}$$

Using Eq. 2.2 and this definition we can write

$$I_c = v \bar{\sigma}(\mathbf{a}) \iint_{V_\tau} N_a N_b \epsilon \, d\mathbf{x} \, d\mathbf{s} \tag{2.5}$$

or

$$\bar{\sigma}(\mathbf{a}) = \frac{I_c}{v} \left[\iint_{V_\tau} N_a N_b \epsilon \, d\mathbf{s} \, d\mathbf{x} \right]^{-1} \tag{2.6}$$

It is an essential part of the experiment whose object is to measure $\bar{\sigma}(\mathbf{a})$ to evaluate the integral

$$G(\mathbf{a}) \equiv \iint_{V_\tau} N_a N_b \epsilon \, d\mathbf{s} \, d\mathbf{x} \tag{2.7}$$

which appears in brackets in Eq. 2.6. Many investigators have not devoted sufficient attention to the proper evaluation of the integral.

2.3 DIFFERENTIAL CROSS SECTION MEASUREMENTS

In a differential cross section measurement, an attempt is usually made to design the apparatus so that $\epsilon = 0$ for all values of $\mathbf{s}$, except those in a small neighborhood of a particular value $\mathbf{s}_0$. The value of $\mathbf{s}_0$ is a function of the apparatus configuration $\mathbf{a}$. If $\sigma(\mathbf{s})$ is essentially constant in the selected domain of s values surrounding $\mathbf{s}_0$, then the average differential cross section $\bar{\sigma}(\mathbf{a})$ defined in Eq. 2.6 is equal to $\sigma(\mathbf{s}_0)$ and one may write

$$\sigma(\mathbf{s}_0) = \frac{I_c}{v} \left[\iint_{V_T} N_a N_b \epsilon \, d\mathbf{s} \, d\mathbf{x} \right]^{-1} \tag{2.8}$$

When this equation is applied in the experimental determination of a differential cross section, the quantities from which the cross section is derived are I_c, v, $N_a(\mathbf{x})$, $N_b(\mathbf{x})$, and $\epsilon(\mathbf{x}, \mathbf{a}, \mathbf{s})$. The accuracy of the result depends on the accuracy with which these quantities are determined.

Usually $\epsilon(\mathbf{x}, \mathbf{a}, \mathbf{s})$ is determined by calculations based on the geometrical and electrical parameters of the product selector and property analyzer. Quantities $N_a(\mathbf{x})$ and $N_b(\mathbf{x})$ are determined in a variety of ways depending on special circumstances of the measurement and the types of reactants involved.

2.4   TOTAL CROSS SECTION MEASUREMENTS

In conducting a total cross section measurement an investigator usually attempts to design his apparatus so that ϵ is independent of $\mathbf{s}$. When this condition holds, the measurement does not discriminate against the property of the product whose production cross section is to be measured. In that case Eq. 2.2 can be written in the form

$$I_c = v \int_V N_a(\mathbf{x}) N_b(\mathbf{x}) \epsilon(\mathbf{x}, \mathbf{a}) \, dx \int_T \sigma(\mathbf{s}) \, d\mathbf{s} \tag{2.9}$$

If we define the total cross section as

$$\sigma_t \equiv \int_T \sigma(\mathbf{s}) \, d\mathbf{s} \tag{2.10}$$

we have

$$\sigma_t = \frac{I_c}{v} \left[\int_V N_a N_b \epsilon \, dx \right]^{-1} \tag{2.11}$$

When this equation is applied, the quantities from which the cross section is derived are I_c, v, $N_a(\mathbf{x})$, $N_b(\mathbf{x})$, and $\epsilon(\mathbf{x}, \mathbf{a})$. The accuracy of the result depends on the accuracy with which these critical quantities are determined.

2.5 THE INTERMEDIATE CASE: CROSS SECTIONS THAT ARE NEITHER TOTAL NOR DIFFERENTIAL

Sometimes a measurement is neither a total cross section measurement nor a differential measurement because $\epsilon(\mathbf{x}, \mathbf{a}, \mathbf{s})$ is neither independent of $\mathbf{s}$, nor equal to zero, except in a small domain of $\mathbf{s}$ values. In such a case, one measures an average value of $\sigma(\mathbf{s})$ weighted according to the function $F(\mathbf{a}, \mathbf{s})$ as defined in Eq. (2.3). If $F(\mathbf{a}, \mathbf{s})$ is not well-defined, the measurement is meaningless. If $F(\mathbf{a}, \mathbf{s})$ is well-defined, the measurement has a specific meaning, even if it is cumbersome to compare it with other experimental results for which F is not the same.

An approximate comparison of a measurement of the intermediate class with a complete set of differential measurements $\sigma(\mathbf{s})$ is possible by numerically integrating the differential experimental results with the experimental weighting factor $F(\mathbf{a}, \mathbf{s})$ under the integral. The same procedure can be used to compare a measurement of the intermediate type with a theoretically predicted differential cross section $\sigma(\mathbf{s})$.

A common example of the intermediate class of cross section is one that is obtained experimentally when a certain class of particles is excluded from a would-be total cross section measurement. Even though the excluded class may be thought to be "small," it is important to render the measurement meaningful by specifying the excluded class. The function $F(\mathbf{a}, \mathbf{s})$ provides a vehicle for this specification.

CHAPTER 3

EXPERIMENTAL METHODS

3.1 BEAM–GAS CONFIGURATION

A special collision arrangement that has been used far more than any other is the so-called beam-gas arrangement in which a beam of particles of type b of well-defined energy passes through a cell containing a gas of particles of type a in thermal equilibrium with the cell walls. In order that the relative velocity of the reactants be well-defined it is necessary that the thermal velocity spread of the gas particles a give rise to a very narrow spread in the magnitude and direction of the relative velocity $\mathbf{v} = \mathbf{v}_b - \mathbf{v}_a$ of the two reactants.

Usually the beam enters and leaves the gas cell through small entrance and exit apertures. The exit aperture size and position are adjusted to transmit all beam particles that enter the cell through the entrance aperture in order that the total current of beam particles which passes through the cell can be accurately determined by a suitable beam detector located downstream from the interaction region. The type a particles are admitted to the cell continually through a leak valve and flow outward into the rest of the system through the apertures. A diffusion or ion pump is employed to reduce the density N_a to an insignificant level in the adjoining chambers through which the beam particles must pass before and after traversing the gas cell.

The density N_a within the cell is usually so low that any product signal I_c from the interaction region is directly proportional to N_a so as to ensure that multiple collisions of beam particles are very rare compared to single collisions. The number density N_a in the cell is most often determined by a pressure or density measurement inside the cell at a fixed monitoring point some distance from the apertures. The value of $N_a(\mathbf{x})$ at all points $\mathbf{x}$ in the interaction region is then assumed to be equal to that at the fixed point or is inferred from molecular flow considerations based on the entire flow system geometry. The interaction region, defined as the region in which the product $N_a(\mathbf{x})N_b(\mathbf{x}) \neq 0$, consists of all points $\mathbf{x}$ inside the gas cell in which the beam particle density $N_b(\mathbf{x}) \neq 0$ and extends outside the apertures where $N_a(\mathbf{x})$ rapidly decreases to a negligible value. The interaction region effectively terminates a few aperture diameters outside the openings in most arrangements.

The general formula for the cross section given by Eq. 2.6 can be simplified by expressing $N_a(\mathbf{x})$ and $N_b(\mathbf{x})$ as follows:

$$N_a(\mathbf{x}) = N_a(\mathbf{x}_0)f_a(\mathbf{x}) \tag{3.1}$$

$$N_b(\mathbf{x}) = \frac{I_b}{v_b}f_b(\mathbf{x}) \tag{3.2}$$

where $N_a(\mathbf{x}_0)$ is the number density measured at a fixed monitoring point $\mathbf{x}_0$ within the gas cell, $f_a(\mathbf{x})$ is a relative gas density function defined at all points in the interaction region, I_b is the total current of beam particles traversing the collision cell, $f_b(\mathbf{x})$ is a relative beam particle density profile, and v_b is the beam particle velocity. Using Eqs. 3.1 and 3.2 in conjunction with Eq. 2.6, we have

$$\bar{\sigma}(\mathbf{a}) = \frac{I_c}{I_b N_a(\mathbf{x}_0)}\left[\iint\limits_{\tau V} f_a(\mathbf{x})f_b(\mathbf{x})\epsilon(\mathbf{x}, \mathbf{a}, \mathbf{s})\, d\mathbf{x}\, d\mathbf{s}\right]^{-1} \tag{3.3}$$

It is often possible to design a beam-gas apparatus so that the following conditions are fulfilled: (1) f_a is equal to a constant k everywhere that the product $\epsilon f_b \neq 0$, and (2) ϵ is constant across any plane normal to the beam velocity vector $\mathbf{v}_b$. Then Eq. 3.3 simplifies to the form

$$\bar{\sigma}(\mathbf{a}) = \frac{I_c}{I_b k N_a(\mathbf{x}_0)}\left[\iint\limits_{\tau\, z} \epsilon(z, \mathbf{a}, \mathbf{s})\, dz\, d\mathbf{s}\right]^{-1} \tag{3.4}$$

where z is the coordinate of axial position parallel to the beam velocity. In deriving Eq. 3.4, we have made use of the fact that the integral

$$\int\limits_{s} N_b(\mathbf{x})v_b\, dx\, dy \tag{3.5}$$

where s is any surface normal to the beam axis, equals the total beam current I_b passing through the entrance aperture. This relation implies

$$\int\limits_{s} f_b(\mathbf{x})\, dx\, dy = 1 \tag{3.6}$$

when f_b satisfies Eq. 3.2. By exercising care in designing his apparatus, an investigator often can reduce $\epsilon(z, \mathbf{a}, \mathbf{s})$ to a form that can be easily calculated with high precision.

The expressions for the differential and total cross section corresponding to Eqs. 2.8 and 2.11 for beam-gas experiments satisfying the preceding conditions on f_a and ϵ are, respectively,

$$\sigma(\mathbf{s}_0) = \frac{I_c}{I_b k N_a(\mathbf{x}_0)} \left[\iint_{\tau z} \epsilon \, dz \, d\mathbf{s} \right]^{-1} \qquad (3.7)$$

and

$$\sigma_t = \frac{I_c}{I_b k N_a(\mathbf{x}_0)} \left[\int_z \epsilon \, dz \right]^{-1} \qquad (3.8)$$

In dissociation experiments either a or b is a molecule and sometimes both are molecules. Dissociation in collisions between reactants a and b in the gas cell is usually detected by the observation of a flux of dissociation fragments emerging from the collision chamber. The most common test for distinguishing beam-gas reaction products from various background effects, such as production at slit edges or on collision chamber walls, is to look for a growth of the detected fragment flux with increase in collision chamber pressure. Pressures are generally maintained in the range 10^{-6} to 10^{-4} torr, or sufficiently low that a negligible fraction of the type b particles undergoes more than one collision in traversing the gas cell. When this condition holds and if the fragments do not suffer attenuation in the gas prior to reaching the collector, the growth of fragment production is proportional to the pressure.

For quantitative measurements of the cross section for a dissociative collision it is necessary to have suitable means of measuring the collision chamber pressure, the beam particle current, and the collision product particle current. It is essential to define the geometry so that, in effect, the function $\epsilon(\mathbf{x}, \mathbf{a}, \mathbf{s})$ can be calculated and communicated along with the measurements. When a differential measurement is made on the production of a product having a selected velocity and/or emission solid angle, it is necessary to provide a good definition of the resolution in regard to the selected quantities. These tasks are accomplished through the use of precise collimators and electrostatic and magnetic deflectors having known ion transmission characteristics.

Mass spectrometry plays an important role in the preparation of the beam particles as well as in the identification of the fragments in most dissociation measurements. We refer the reader to the treatise by Ewald and Hintenberger[5] for information on ion sources, magnetic and electrostatic focusing, and detection.

The majority of the dissociation experiments have been performed using ions as reactant b and neutrals as reactant a. A gas target usually

comprises neutral atoms or molecules even at the high tempratures sometimes used[6] to induce dissociation of the gas. The beam intensities are
always low enough so that no significant portion of the collision chamber
gas is excited or ionized.

We now proceed to discuss the various measurement methods in more
detail. It is convenient to partition the beam-gas measurements into
two major subdivisions that we call projectile molecule dissociation measurements and target molecule dissociation measurements. These are discussed in Sections 3.2 and 3.3. In Section 3.4 we touch briefly on the
crossed-beam method for which relatively little experience exists in the
field of dissociation studies.

3.2 PROJECTILE MOLECULE DISSOCIATION
MEASUREMENTS

The first major class of dissociation measurements is that in which
a projectile molecule or molecule ion is directed into a gas target and
the dissociation fragments of this projectile are analyzed on emergence.
In most of the literature describing results of this method, both the
projectiles and observed fragments were ions. This situation is relatively
easy to manage since mass-to-charge ratio analysis can be applied
to both the projectiles and fragments for positive species identification.
We defer a discussion of the neutral projectile dissociation measurements
to the end of this section. We discuss also at that point the detection
of neutral projectile fragments.

The apparatus forms employed in projectile dissociation measurements
are extremely varied in details, but a few main types are dominant.
The oldest is the so-called Aston band or "fractional mass peak"
method.[7-13] This method employs an ordinary mass spectrometer consisting of the following elements: an ion source, ion accelerator, ion drift
space, magnetic analyzer, and ion detector. In this type of apparatus,
molecular ions emerging from the accelerator section undergo dissociative
collisions with gas molecules or atoms in the drift space between the
accelerator and magnetic deflector. The dissociation fragments, proceeding usually with a velocity almost equal to that of the primary ion,
have a different momentum-to-charge ratio from that of the primary
ions and consequently require a different magnetic deflection field to
bring them into the detector. The field required to accomplish this is
the same as that which would be required to select particles of mass-to-
charge ratio $(m/e)^*$, moving directly from the ion source to the detector

without a collision. The value of $(m/e)^*$, usually a fraction, is

$$\left(\frac{m}{e}\right)^* = \left(\frac{m}{e}\right)_f^2 \left(\frac{m}{e}\right)_i^{-1}$$

where $(m/e)_f$ is the dissociation fragment mass-to-charge ratio and $(m/e)_i$ is the parent molecule ion mass-to-charge ratio.[14] The ion current received at the collector, plotted as a function of the magnetic field, is called an Aston band. Usually Aston bands are broader than ordinary mass peaks not involving drift tube dissociation because a spread in the direction and velocity of the fragments relative to the parent ion is introduced in the dissociation process. Directional spread of the fragments introduces a broadening of the band unless the geometry of the magnetic deflector is such that dissociation fragments originating all along the drift space are sharply focused at the detector aperture, and velocity spread introduces a spread in the deflection angles in the magnetic deflector so as to broaden the ion impact area at the collector slit. Dissociation cross sections may be determined with apparatus of this type when due regard is given to the critical features to be discussed in Chapter 5. A precaution peculiar to the Aston band method that must be observed in quantitative work is the consideration of all the possible values of $(m/e)_i$ of the ions which can emerge from the accelerator and all of the possible $(m/e)^*$ values associated with the possible dissociation fragments of the parent ions. In some cases ambiguities in fragment identification can arise because Aston bands from two or more processes overlap. An example is discussed by Potapov.[15]

Beynon et al.[16] have derived a useful expression for the width of an Aston band produced in two-fragment dissociation in terms of the amount of energy that is converted from internal energy of the dissociating molecule to kinetic energy of the two fragments. With a change in notation to provide consistency with the terms in the foregoing equation, Beynon's formula is

$$\left(\frac{m}{e}\right)^* = \left(\frac{m}{e}\right)_f^2 \left(\frac{m}{e}\right)_i^{-1} \left[1 \pm \sqrt{\frac{\mu T}{e_i V}}\right]^2$$

where the two values of $(m/e)^*$ corresponding to the two signs in the bracket denote the extreme apparent mass deviations from the center of the Aston band, μ = the ratio of the mass of the observed fragments to the mass of the other fragment, T = the total kinetic energy of the fragments relative to their mass center, V = the projectile acceleration voltage, and e_i = the projectile charge. It is interesting to note that when $T/e_i V = 0.001$ and $\mu = \frac{1}{2}$, the two values of $(m/e)^*$ calculated for the ends of the Aston band are $\pm 4.4\%$ above and below the band center. This expression was derived for the case of spontaneous dissociation, but it should also hold for collision-induced dissociation provided that the collision process does not significantly affect the velocity of the center of mass of the fragments.

A variant of the Aston band method has been employed by Harris[17] in conjunction with a Dempster type-mass spectrometer. This configuration seems to be less satisfactory than those employing a straight line drift space, since the effective path length is not well defined. Improvements in geometry might alleviate the defects in this apparatus.

Makov and co-workers[18] have studied the dissociation of H_2^+ projectiles in an apparatus employing a single magnetic analyzer to select the projectile and to analyze the H^+ fragment production. This apparatus appears to have satisfactory properties for the analysis of simple dissociation processes involving one ion fragment species only, but the double mass spectrometer systems, to be discussed below, are preferred for complex molecule studies.

Henglein[19,20] has employed a "parabolic" mass spectrometer in conjunction with photographic plates to obtain qualitative data on dissociation fragments of several molecular ions. This method has not yielded quantitative data on dissociation cross sections or relative fragment yields, but it is very useful for examining the gross characteristics of fragmentation patterns and energy distributions quickly and easily. The method has the advantage of recording all fragments at once for all species emergent from the ion source.

Another major type of dissociation apparatus used for the study of projectile molecule dissociation employs a double mass spectrometer consisting of the following sequence of elements: ion source, ion accelerator, magnetic deflector, collision chamber magnetic deflector, and detector. This type of instrument[21-25] avoids the ambiguities in identification of mass peaks which sometimes occur in the Aston band type of instrument. Some models of this design are superior in that the collision region is of short and well-defined length. A short collision region approximates a point source of dissociation fragments so as to assist in focusing the collision products onto a detector aperture. Accurate definition of the length of the collision region assists in the accurate calculation of cross sections.

A variation of the above design substitutes an electrostatic deflector for the second magnetic deflector.[26-30] An electrostatic deflector performs a separation of beam-ion dissociation fragments according to energy-to-charge ratio instead of momentum-to-charge ratio. In the forms that have been invoked in existing measurements, there are no apparent advantages of one system over the other. If one were to attempt to achieve high-quality two-dimensional focusing of fragment beams, it is conceivable that electrostatic deflection might be chosen for economy.

Yet another variation of double mass spectrometer[31] for fast fragment analysis employs both an electrostatic and magnetic deflector in sequence to analyze projectile dissociation fragments emergent from the collision chamber. This type of instrument is quite complicated to analyze and to operate and is not especially desirable unless the complexity is actually required, as when one wishes to deal with fragments having broad energy spreads and closely spaced mass-to-charge rations. In some cases it is simply not possible to separate adjacent chemical species with an electrostatic or magnetic analyzer alone.

Hunt and co-workers[32,33] and Ferguson and co-workers[34] have used time-of-flight mass spectrometers to study dissociation fragments of fast projectile molecules produced by collisions in the instrument drift tube. By application of a retarding field near the end of the flight path, peaks due to neutral species and fragment ions are separated from their parent ion peaks. All of the peaks can be displayed in the same panoramic mass spectrum.[34] Only qualitative results of this method have been presented. The advantages of time-of-flight technique over the magnetic or electrostatic analyzer techniques are not apparent.

Several studies of dissociation at MeV energies have been conducted by observing the attenuation of a beam of molecular ions in a cyclotron while the ions were being accelerated under cyclotron resonance conditions.[35-37] These experiments measure the probability that an ion changes its mass-to-charge ratio by a gas collision in following a regular spiral path between two selected radii. The analysis of the data to obtain a dissociation cross section requires the assumptions that attenuation is due entirely to dissociation, and that the energy change per revolution is known. The calculation of the ion total path length between two radii rests on a determination of the energy change per revolution. Two of the experiments[35,37] involved an additional assumption that the relative cross section varied in proportion to the inverse of the ion energy. The third study[36] afforded a means of checking this assumption, since it provided a measure of the neutral fragment production as a function of orbit radius.

In some instances dissociation of a molecular ion can be inferred from the attenuation of a beam of ions as the beam passes through a gas cell. Such an inference can be reached when total absorption cross section is measured and when that cross section dominates, or can be corrected for, nondissociative modes of beam attenuation include scattering and change of charge through electron capture or loss without dissociation.

Examples of applications of the attenuation method can be found in the data index tables under the label P5.

Several investigations have been directed to the precise determination of projectile dissociation fragment energy distributions. Examples are the work of Caudano and Delfosse,[38] Rourke, Shefield, and Davis,[31] and Durup, Fournier, and Pham,[39] in which fragments of fast molecular ions were highly collimated and then energy-analyzed.

Studies incorporating correlated angular and energy distribution measurements have been conducted by Gibson et al.[40-43] and Vogler et al.[44] These studies employ refined geometrical features, but incorporate no basically novel apparatus elements. Energy and angular distribution data are especially labeled in the index table by the symbols P2 and P3, respectively. These studies have greatly illuminated the finer details of the dissociation mechanism for diatomic molecules because they assist in establishing the postcollision state of the molecules leading to their breakup. They also provide important design information for those who wish to construct apparatus for total cross section measurements.

Several investigators have utilized electrostatic retardation of product ions to study fragment energy distributions. This method has been applied to projectile fragments[12] and to fragments that may have included an indistinguishable mixture of projectile and gas target dissociation fragments.[45] Indistinguishability of fragments on the basis of energy analysis is a possibility that can occur when the collision energy is of the same order of magnitude as the kinetic energy with which the gas molecule dissociation fragments are formed.

A number of investigators have observed dependence of the collisional dissociation probability of a molecular ion on the operating parameters of the ion source or on the parent species from which a molecular beam ion originates. (The same molecular ion may be produced, in general, from any molecular gas that can form the ion in question by splitting under electron impact or by a sequence of events involving ionization followed by ion molecule reactions.) Many of these studies are described in Chapter 6. Among the easiest to interpret are those that, following the early work of Tunitskii et al.,[7] have employed ion sources in which the electron bombardment energy was known and controllable, and in which collisions of the ions in the ion source subsequent to formation were made negligible by the use of a low ion source pressure. Investigations of this type and others related to ion source effects are labeled by the symbols P1, P2, or P3, followed by various letters a through d in the data index table. Some of these studies have been ill-defined with respect to specification of the velocity and angle discrimination imposed by the collector geometry on the dissociation fragments observed so that the cross sections belong to the intermediate case discussed in Section 2.5. An appreciation of the importance of this specification of the accepted products is reflected in the recent work of Durup et al.,[39] wherein precise geometrical details of the slit system were provided.

In the greater part of the dissociation literature, attention has been concentrated on the probability of production of a particular dissociation product whose mass-to-charge ratio was identified by means already discussed. In many studies, several of the possible fragments from a given parent species were observed (see column 4 of the data index table in Chapter 4 for examples). In a very few instances it is possible to obtain information on the relative probabilities of dissociation by particular fragmentation modes. Partial or complete separation has been accomplished in the work of Sweetman,[21,46] Riviere and Sweetman,[47] Guidini,[48] and Berkner et al.[37] for the H_2^+ ion. The techniques used are limited to those energies at which single particle detector response and wide aperture angles are possible. These conditions imply projectile energies greater than about 20 keV. From these measurements separate cross sections have been obtained for the following breakup modes of the H_2^+ ion:

$$H_2^+ \rightarrow H^+ + H$$
$$H_2^+ \rightarrow H^+ + H^+$$
$$H_2^+ \rightarrow H \ + H$$

In some of the measurements the cross section for the mode

$$H_2^+ \rightarrow H_2$$

is included as an unresolved fraction of the total cross section for the pair of modes

$$H_2^+ \rightarrow H + H$$
$$H_2^+ \rightarrow H_2$$

However, even this mixture can be resolved by the use of narrow detector slits.[21] A sufficiently narrow slit discriminates against the upper channel of the last pair because the two atoms in the pair $H + H$ usually have sufficient relative velocity to strike the detector at points having a macroscopic separation.

A less direct method of determining breakup mode cross sections of H_2^+ ions has been used by Ropke and Spehl.[49] This method involves a study of the growth of the neutral H atom signal with collision chamber pressure and requires a knowledge of the cross section for the neutralization of protons. This technique is applicable only at collision velocities $> 5 \times 10^8$ cm sec^{-1}, where the last two processes listed above can be ignored.

In the case of triatomic and other polyatomic ions, where three particle breakup modes are possible, the experimental difficulties attendant to mode resolution are large. An upper limit to the cross section for any one mode is of course obtained by a measurement of the cross section for production of any one fragment resulting from the mode. The lowest upper bound for a given mode obtainable in this way is the lowest of the cross sections for the individual fragments of the mode.

In enumerating the possible modes of dissociation it is important to recall that charge exchange or electron loss may occur from the parent molecule leading to a set of fragments whose total number of bound electrons is not the same as that of the primary reactants. This is illustrated in the H_2^+ mode resolution work.

In a number of investigations employing mass spectrometers, Aston bands or other evidence of dissociation of ions in flight has been obtained which fails to pass the test of proportionality of dissociation product signal to pressure of gas in the drift space. Instead, it is sometimes observed that the yield versus pressure curve has a finite positive intercept at zero pressure. This behavior has been invoked by several investigators as evidence for the spontaneous dissociation of molecular ions as they proceed down the drift tube. Such behavior and interpretation was given by Kupriyanov[50] to observations of CO^{2+} ions. In this case a lifetime of $\sim 4 \times 10^{-4}$ sec was inferred from the fraction of the ions which decayed in a known distance at a known velocity. Apparently, a quantum mechanical tunneling process is responsible for the spontaneous dissociation in this instance. Although this type of process is distinct from the process of collision-induced dissociation, its presence could lead to erroneous results in some experiments, especially those in which pressure dependences are not observed in every cross section determination. It is also possible that polyatomic dissociation products of collisional dissociation may sometimes decay after leaving the collision region but before arrival at the detector. These events proceeding unrecognized could lead to errors in cross section measurements and/or unexplained discrepancies among experiments with different drift distances between the collision region and the detector.

In cases where spontaneous dissociation does indeed result from tunneling, one would expect the lifetime of the unstable species to be strongly dependent on the vibrational and rotational states of the species and therefore on the exact conditions under which the species is formed.

Experimental signal backgrounds due to spontaneous dissociation consequently might vary strongly with minor inadvertent changes in experimental conditions.

A portion of the literature dealing with spontaneous dissociation is represented by refs.[8,9,14,16,51-62].

The fragments of projectile dissociation in which the projectile energy is much greater than molecular bond energy tend to appear with nearly the same direction and velocity as the beam ions. Actually there is always a finite scatter of the direction and velocity of the fragments, but both spreads are sometimes small enough when the beam energy is large so that simple ion optics can be designed to receive essentially all fragments of each mass-to-charge ratio without interference between species owing to overlap of the deflected beams emergent from the second analyzer. The absolute spread in direction and the percentage spread in velocity both increase as the projectile velocity decreases. Special care must therefore be used to insure complete fragment collection at lower beam energies. In some measurements incomplete collection probably occurred at energies as high as 10 to 20 keV. Several authors have discussed these effects.[8,24,28,63,64]

In a few dissociation experiments[65-68] an ion beam has been converted to a neutral beam in a charge exchange cell placed directly in front of the collision chamber in order that neutral-neutral collisions may be studied. It is, of course, necessary to remove residual ions from the beam prior to its entry into the collision chamber by means of some device such as an electrostatic or magnetic deflector. It is necessary also to consider the implications of the presence of more than one neutral species in the neutral beam when a molecular ion beam is employed as the incident beam to the gas cell because neutral dissociation fragments may accompany the neutralized molecular ions in the product beam issuing from the exit of the neutralizing chamber.[65]

A few measurements[21,27,48,63,65,69] have determined the yield of neutral dissociation fragments of molecular beam particles. In these studies the total ionization produced in a gas or solid counter by individual neutral dissociation fragments was measured and pulse height was used as a means of discrimination between neutralized undissociated molecular beam ions and neutral dissociation fragments. This method has been applied only to H_2^+ and H_2 but is probably applicable to other species in which the possible neutral fragments have clearly distinguishable pulse heights. Fragment energies greater than a few keV are required for pulse height analysis using presently available detectors. It is not now considered feasible to analyze neutral fragments of the gaseous reactant because the energies of such fragments are almost all lower than a few eV. A further difficulty in the detection and identification of these neutrals is that they have a large ratio of energy spread to mean energy.

3.3 TARGET MOLECULE DISSOCIATION MEASUREMENTS

In this section we describe the methods that have been employed to measure the dissociation of molecules of a gaseous target under bombardment by a beam of ions or neutral particles. The problem of complete collection of fragment ions and the characterization of those that are collected is difficult when the fragments of interest are those of the gaseous species in the collision chamber. These are emitted through a 4π solid angle, are very likely emitted anistropically, and may be expected to have energy spreads comparable to their mean energies.

The most often used type of apparatus for the study of the gas fragments employs a parallel plate electrostatic condenser within the collision chamber.[70-78] An applied voltage accelerates the fragment ions toward one electrode which is supplied with a narrow slit. A mass spectrometer just outside the slit performs a selection of the fragment ions according to mass-to-charge ratio. Some apparatus of this kind employ a condenser with field oriented at right angles to the beam direction[70-73,76-78] and some employ a condenser with field oriented parallel to the beam direction.[74,75]

Slow fragment analyzers that employ electric fields within the collision chamber have the property of deflecting the beam ions and modifying their energy as they traverse the chamber. They also accelerate fragment ions prior to collection by an amount that is a function of the point of ion formation. Both of these effects need to be accounted for in the analysis of the data for purposes of total or differential dissociation fragment yield determinations.

Electric-field ion collectors with contiguous mass spectrometers tend to discriminate against ions that have an initial velocity component perpendicular to the collecting field. All such ions enter the collector slit at an angle deviating from the normal to the collector plate and at a finite distance from the normal projection of the point of origin of the ion on the collector electrode. To insure ion collection without discrimination, the collector slit and ion optics leading through the entire mass spectrometer and detector must be designed so that the entire width of the ion impingement trace on the collector electrode is accommodated and so that the most extreme angles of entry are accommodated. Browning and Gilbody[78] have considered these factors in great detail and have developed apparatus well-suited to the problem.

Another version of slow fragment analyzer employs a field-free collision region[45,79] with a collimator pointed at the beam and oriented to receive fragments in a particular emission angle range. The ions emergent from the collimator are analyzed by means of a mass spectrometer. An especially valuable and flexible example of this type of apparatus is that of Champion, Doverspike, and Bailey[79] in which the orientation of the collimator is variable so that fragment angular distributions can be taken.

At very low beam energies (of the order of 10^2 to 10^3 eV and lower), apparatus of the above description[79] is useful in dealing with the broadened angular distribution of projectile dissociation fragments, as well as the very broad angular spread of the gas dissociation fragments.

A method for studying the dissociation fragments produced by ion bombardment of a molecular gas involving the use of only one mass spectrometer has been developed by Cermak and Herman[80] and applied by Galli et al.,[77] Futrell and Tiernan,[81] and Henglein and Muccini[82] to dissociation studies. This apparatus uses a special ion source in which ions are both produced and accelerated in a single gas-filled chamber. As employed to date, this method does not produce projectile ions of well-defined energy. In addition, great care must be exercised to insure that the projectile ion species is known when it is energetically possible that more than one such species can be produced either by direct electron bombardment of the projectile source gas or by ion molecule reactions of ions in the source region. In general, we feel that a double mass spectrometer—one spectrometer for projectile selection and a second spectrometer for product selection—is superior to the simple single mass spectrometer arrangement for definitive quantitative work.

A time-of-flight mass spectrometer has been employed by Homer and co-workers[83,84] to observe the mass spectra of gas molecules that were dissociated under ion impact. These studies did not utilize a mass-analyzed projectile beam.

The dissociation of a gaseous molecule comprising n-bound electrons can be inferred from the conversion of projectile atomic ions of charge m into ions of charge $m - n$ in single collisions occurring in the gas cell. This type of collision entails the complete stripping of electrons from the gaseous molecule and implies dissociation as a result. This type of measurement has been used primarily where the gas in question is H_2. A few examples can be found in the index table, Chapter 4, where H_2 was converted to two protons by double electron capture of singly charged ions.[85-88] Ordinary beam-gas charge transfer apparatus is useful for this type of measurement if features are incorporated for determining the postcollision charge state of projectile ions.

Often the dissociation of a gas molecule or a projectile molecule in a beam-gas arrangement can be inferred from the observation of characteristic photon emissions from dissociation fragments produced in excited states. Numerous examples of this type of study are to be found in the literature. A recent compilation of heavy particle collision excitation data sources by E. W. Thomas[89] encompasses most of the work. We have not reviewed this type of data in this report, but we wish to recommend it as a source of important information on excited dissociation fragment production. This kind of data can assist in further defining some of the many channels available for fragments produced on collisional dissociation.

3.4 CROSSED-BEAM DISSOCIATION MEASUREMENTS

Three of the studies included in the present review were conducted by means of the crossed-beam method.[75,76,90] In this method both reactants were directed into an interaction region in the form of beams, and dissociation products emergent from the interaction region were observed by suitable instrumentation similar to that used in beam-gas measurements. This method is usually used when neither of the two reactants to be investigated exists in the gaseous form, for example, when both are ions as in the work of Sinda.[90] The use of the crossed-beam technique is accompanied by three difficulties which would ordinarily discourage its use when the beam-gas method is applicable: (1) The rate of reaction in the beam overlap region is usually appreciably smaller than that in a typical beam-gas experiment because available beams contain much lower particle densities than typical gas cells. (2) The density of molecules of the residual gas in the vacuum chamber is usually higher than that in the beams so that large signal backgrounds are present from collisions of both beams with the residual gas. The separation of signal from noise is usually accomplished by beam modulation and phase sensitive detection. (3) The procedures required to determine absolute cross sections entail determination of the particle flux profiles in both beams. These problems have been given special attention by a number of investigators, but we will not dwell further on these problems because of the relatively slight use the technique has found in the study of dissociation collisions. Nevertheless, a very substantial portion of the criteria discussion in Chapter 5 relates to crossed-beam studies as well as beam-gas studies.

CHAPTER 4

EXPERIMENTAL DATA INDEX

In this chapter we present an index to the dissociation literature which classifies the experimental data according to type. Data exist for the 101 molecular species listed in Table 4.1. Also shown in the table is the number of atoms in the molecule, the charge carried by molecule, and a page number which refers to the detailed data listing in Table 4.2. The compound labeled "species" in Table 4.1 is found in the first column of Table 4.2.

Table 4.2 is a detailed guide to the available dissociation literature. It is constructed in such a way as to permit a quick and comprehensive survey of the type of data available. Nine types of information are supplied in the various columns of the table as described in the table caption.

Experimental data designated as erroneous by the author have not been listed in this Table 4.2; however, the source of our information which states that a publication contains erroneous data is given in Appendix I, and the paper is listed in the reference list even though the original author may have condemned the entire data content of his paper.

Experimental data designated by their author as having been superseded by a more accurate or more comprehensive presentation are not listed in this index, but the basis of our information that the data were superseded is listed in Appendix I and the superseded reference is listed in the reference list.

Experimental data from several theses are listed in the data index table.[91-98] When thesis data have been partially or wholly published, the contents of the publications are listed separately in the index table. We have retained the thesis data listings on dissociation in their entirety since the theses almost always contain essential information about the experimental method and error sources and often provide more comprehensive data displays.

26

TABLE 4.1

List of molecular species for which numerical experimental data are available. Symbol n refers to the number of atoms in the molecule. Symbol q refers to the charge of the molecule in electron charge units. The number in parenthesis is the molecular weight rounded off to the nearest integer. The page numbers refer to the location of the expanded data listings in Table 4.2.

n	q	Species	Page
2	2	$CO^{+2}(28)$	32
2	2	$N_2^{+2}(28)$	32
2	2	$SO^{+2}(48)$	32
2	2	$SF^{+2}(51)$	32
2	1	$H_2^+(2)$	33-40
2	1	$HD^+(3)$	40-41
2	1	$D_2^+(4)$	41-42
2	1	$HeH^+(5)$	42
2	1	$CH^+(13)$	42
2	1	$NH^+(15)$	42
2	1	$OH^+(17)$	42
2	1	$C_2^+(24)$	42
2	1	$CO^+(28)$	42-43
2	1	$N_2^+(28)$	43-44
2	1	$N_2^+(29)$	44
2	1	$NO^+(30)$	44
2	1	$O_2^+(32)$	44
2	1	$SO^+(48)$	44
2	1	$SF^+(51)$	44
2	0	$H_2(2)$	45-46
2	0	$CO(28)$	46-48
2	0	$N_2(28)$	48-51
2	0	$NO(30)$	51
2	0	$O_2(32)$	51-52
2	-1	$NaI^-(150)$	52
2	-1	$Sb_2^-(244)$	52
2	-1	$Te_2^-(256)$	52
2	-1	$Bi_2^-(418)$	52
3	2	$CO_2^{+2}(44)$	52
3	2	$SF_2^{+2}(70)$	52
3	1	$H_3^+(3)$	53
3	1	$D_3^+(6)$	53
3	1	$CH_2^+(14)$	53-54

TABLE 4.1

n	q	Species	Page
3	1	$NH_2^+(16)$	54
3	1	$H_2O^+(18)$	55
3	1	$C_2H^+(25)$	55
3	1	$C_3^+(36)$	55
3	1	$CO_2^+(44)$	55
3	1	$N_2O^+(44)$	55
3	1	$SO_2^+(64)$	55
3	0	$H_2O(18)$	55-56
3	0	$H_2S(34)$	56-58
3	0	$CO_2(44)$	58-59
3	0	$N_2O(44)$	59-61
3	0	$SO_2(64)$	61
3	0	$CS_2(76)$	61
3	-1	$NaI_2^-(277)$	61
3	-1	$Sb_3^-(366)$	61
4	2	$SF_3^{+2}(89)$	61
4	1	$CH_3^+(15)$	61-62
4	1	$NH_3^+(17)$	62
4	1	$C_2H_2^+(26)$	62-63
4	1	$C_3H^+(37)$	63
4	0	$NH_3(17)$	63-65
4	0	$C_2H_2(26)$	65-69
4	0	$AsH_3(78)$	69
5	2	$SF_4^{+2}(108)$	70
5	2	$C_2H_3^{+2}(27)$	70
5	1	$CH_4^+(16)$	70-71
5	1	$CD_4^+(20)$	71
5	1	$C_2H_3^+(27)$	71
5	1	$C_3H_2^+(38)$	71
5	1	$HCOOH^+(46)$	71
5	1	$HCOOD^+(47)$	71
5	1	$DCOOH^+(47)$	71
5	0	$CH_4(16)$	72-77
5	0	$CD_4(20)$	77
5	0	$CH_3Cl(50)$	77-78
5	0	$CCl_3F(136)$	78
5	0	$CH_3I(142)$	78
5	0	$CCl_4(152)$	78

TABLE 4.1

n	q	Species	Page
6	1	$CD_5^+(22)$	78
6	1	$C_2H_4^+(28)$	78-79
6	1	$C_3H_3^+(39)$	79
6	0	$C_2H_4(28)$	79-82
6	0	$CH_2CD_2(30)$	82-83
6	0	$N_2H_4(32)$	83
6	0	$CH_3OH(32)$	83-85
7	1	$C_2H_5^+(29)$	85
7	1	$C_3H_4^+(40)$	85
7	0	$C_2H_5(29)$	85
7	0	$C_2H_3D_2(31)$	85-86
7	0	$C_2D_3H_2(32)$	86
7	0	$CH_3NH_2(31)$	86
8	1	$C_3H_5^+(41)$	87
8	0	$C_2H_6(30)$	87-89
8	0	$C_2D_6(36)$	89-90
9	0	$C_3H_6(42)$	90-91
9	0	$C_2H_5OH(46)$	91-93
10	1	$C_3H_7^+(43)$	93
10	0	$C_3H_6O(58)$	93
11	1	$C_3H_8^+(44)$	94
11	0	$C_3H_8(44)$	94-95
12	0	$C_4H_8(56)$	95
		buten 1	95
12	0	$C_4H_8(56)$	95
		i-buten	95
12	0	$n\text{-}C_3H_7OH(60)$	95-97
14	0	$C_4\dot{H}_{10}(58)$	97-99
14	0	$B_5H_9(64)$	99

The columns in Table 4.2 contain data defined as follows:

Column 1. Molecule observed to dissociate. The mass of the molecule in atomic mass units rounded off to the nearest integer is shown in parenthesis. The charge state of the molecule is written as a superscript. Charges of +1 are indicated by a + sign only. Other charges are indicated by a sign and a number. The unit of charge is the charge of the electron.

Column 2. Collision partner with which molecule in Column 1 collides to induce the observed dissociation. The mass and charge of the collision partner are designated as in Column 1. In a few papers electron impact dissociation data as well as heavy particle induced dissociation data were obtained. Electron impact data are indicated by the symbol e in Column 2.

Column 3. Total energy of the collision partners in the laboratory coordinate system in keV units. In *P*-type measurements the particle in Column 1 possesses most of the collision energy. In *T*-type measurements the particle in Column 2 possesses most of the collision energy. Column 5 designates the *P* or *T* classification.

Column 4. Mass(es) of observed dissociation fragments in atomic mass units rounded off to the nearest integer. A zero superscript designates a neutral fragment; no superscript designates a fragment of charge +1; other charges are indicated by a sign and a number.

Column 5. Measurement classification. Symbols used in this column are defined as follows:

P	Projectile molecule dissociation.
T	Target molecule dissociation.
1	Total cross section for the production of a particular dissociation fragment.
2	Angular distribution of a particular fragment.
3	Energy distribution of a particular fragment.
4	Cross section for dissociation by a particular fragmentation mode.
5	Cross section for charge transfer plus cross section for dissociation, sometimes called attenuation cross section
6	Total dissociation cross section: sum of partial cross sections for all possible dissociation modes.
a	Dependence of measurement on ion source electron energy.
b	Dependence of measurement on the collision process in which the projectile is formed.
c	Dependence of cross section on gross ion source operating parameters.
d	Dependence of cross section on type of ion source with other system elements unaltered.
e	Quantities are deduced for the center-of-mass system of the dissociating molecule.

Column 6. Data Classification.

A	Absolute. Independent of other cross sections.
NS	Normalized to agree with another measurement of *same* process.
ND	Normalized so that the cross section for a *different* process measured with the same apparatus agrees with a known value for that different process.
NT	Normalized to a theoretical result.
R	Relative. Cross section given in arbitrary units.

Column 7. Number of reference containing data as listed in Reference section.

Column 8. Type of data presentation in original paper, Figure, Table, or Text.

Column 9. Page of data location in original paper.

The lines of the data index table are ordered vertically according to the following rules stated in order of priority. (All of those lines for which property 1 is the same are listed consecutively and are ordered according to property 2. All those lines for which properties 1 and 2 are the same are listed consecutively and are ordered according to property 3, etc.) 1. Number of atoms in molecule undergoing dissociation $= n$. 2. Charge of molecule in order of decreasing charge from positive to negative $= q$. 3. Mass of molecule in order of increasing mass. 4. Collision partner characteristics in order (1) (2) (3). 5. Measurement classification. 6. Energy of collision.

TABLE 4.2

1	2	3	4	5	6	7	8	9
$n=2$	$q=2$							
$CO^{+2}(28)$	He(4)	5.1	12	P1	R	99	Tab.7	1139
$CO^{+2}(28)$	He(4)	30-70	16	P3	R	31	Tab.4	198
$CO^{+2}(28)$	He(4)	30-70	12^{+2}	P3	R	31	Tab.4	198
$CO^{+2}(28)$	He(4)	54.8	16	P3	R	31	Fig.6	197
$CO^{+2}(28)$	Ne(20)	5.1	12	P1	R	99	Tab.7	1139
$CO^{+2}(28)$	Ne(20)	5.6	12,16	P4	A	50	Tab.	662
$CO^{+2}(28)$	Ar(40)	5.1	12	P1	R	99	Tab.7	1139
$CO^{+2}(28)$	Kr(84)	5.1	12	P1	R	99	Tab.7	1139
$CO^{+2}(28)$	$H_2(2)$	5.1	12	P1	R	99	Tab.7	1139
$CO^{+2}(28)$	$D_2(4)$	5.1	12	P1	R	99	Tab.7	1139
$CO^{+2}(28)$	$N_2(28)$	5.1	12	P1	R	99	Tab.7	1139
$CO^{+2}(28)$	CO(28)	1.0	12	P1	ND	17	Text	2270
						93	Text	2133
$CO^{+2}(28)$	CO(28)	5.1	12	P1	R	99	Tab.7	1139
$CO^{+2}(28)$	Air(~30)	5.6	12,16	P4	A	50	Tab.	662
$CO^{+2}(28)$	?	?	12,16	P3	R	50	Fig.4	663
$N_2^{+2}(28)$	He(4)	30-70	14	P3	R	31	Tab.4	198
$N_2^{+2}(29)$	$N_2(29)$	2-3.2	14,15	P1	A	12	Fig.4	363
$SO^{+2}(48)$	He(4)	30-70	16	P3	R	31	Tab.4	198
$SF^{+2}(51)$	He(4)	30-70	19	P3	R	31	Tab.4	198

TABLE 4.2

1	2	3	4	5	6	7	8	9
n=2	q=1							
$H_2^+(2)$	$N_2^+(28)$	100-250	1,1°,2°	P1	A	90	Fig.1	2
$H_2^+(2)$	H(1)	3-115	1	P1	ND	6	Tab.1	183
$H_2^+(2)$	H(1)	10	1	P2	R	6	Text	183
$H_2^+(2)$	He(4)	2-50	1	P1	ND	25	Fig.5	68
$H_2^+(2)$	He(4)	2-50	1,1⁻	P1	A	91	Tab.13	168
$H_2^+(2)$	He(4)	10-180	1	P1	A	100	Fig.3	818
$H_2^+(2)$	He(4)	100-200	1,1°	P1	A	26	Tab.3	390
$H_2^+(2)$	He(4)	200-1200	1,1°	P1,	A	29	Fig.4	25
$H_2^+(2)$	He(4)	500-1200	1,1°	P1	A	29	Tab.	26
$H_2^+(2)$	He(4)	3	1	P2	R	43	Fig.3	
$H_2^+(2)$	He(4)	10	1	P2,3e	R	42	Fig.1,2	634-635
$H_2^+(2)$	He(4)	10	1	P2,3e	R	43	Fig.2	
$H_2^+(2)$	He(4)	3	1	P3e	R	43	Fig.3	
$H_2^+(2)$	He(4)	3,5,10	1	P3e	R	43	Fig.4	
$H_2^+(2)$	He(4)	10	1	P3,	R	40	Fig.7abc	268
$H_2^+(2)$	He(4)	30-250	1,1°,2°	P4	A	48	Fig.1	
$H_2^+(2)$	He(4)	40-880	1°,2°	P4	A	46	Dwg.3	
$H_2^+(2)$	He(4)	40-3000	1,1°	P4	A	46	Dwg.1	
$H_2^+(2)$	He(4)	44-3000	1	P4	A	46	Dwg.2	
$H_2^+(2)$	He(4)	20,900	1,1°	P4	A	37	Tab.1	146
$H_2^+(2)$	He(4)	10000-65000	none	P5	A	37	Fig.4	

TABLE 4.2

1	2	3	4	5	6	7	8	9
$H_2^+(2)$	He(4)	100-200	1,1°	P6	A	26	Tab.3	390
$H_2^+(2)$	He(4)	200-1200	1,1°	P6	A	29	Fig.4	25
$H_2^+(2)$	Ne(20	2-50	1	P1	ND	25	Fig.5	68
$H_2^+(2)$	Ne(20)	2-50	1,1⁻	P1	A	91	Tab.13	168
$H_2^+(2)$	Ne(20)	3.8	1	P1b	A	101	Tab.	1223
$H_2^+(2)$	Ne(20)	3.8	1	P1ab	A	102	Fig.1	923
$H_2^+(2)$	Ne(20)	5-29	1	P1	A	23	Fig.11	
$H_2^+(2)$	Ne(20)	200	1,1°	P1	A	26	Tab.1	389
$H_2^+(2)$	Ne(20)	10	1	P2,3e	R	43	Fig.2	
$H_2^+(2)$	Ne(20)	10	1	P2,3e	R	42	Fig.2	635
$H_2^+(2)$	Ne(20)	10	1	P3e	R	43	Fig.4	
$H_2^+(2)$	Ne(20)	20.4	1	P3	R	44	Fig.7,8,9	350-352
$H_2^+(2)$	Ar(40)	.012-.135	1¹	P1	A	103	Fig.1b	803
$H_2^+(2)$	Ar(40)	.012-.135	1	P1	A	96	Tab.2	33
$H_2^+(2)$	Ar(40)	2-50	1	P1	ND	25	Fig.5	68
$H_2^+(2)$	Ar(40)	2-50	1,1⁻	P1	A	91	Tab.13	168
$H_2^+(2)$	Ar(40)	5-29	1	P1	A	23	Fig.11	
$H_2^+(2)$	Ar(40)	10-180	1	P1	A	100	Fig.4	
$H_2^+(2)$	Ar(40)	150,200	1,1°	P1	A	26	Tab.3	390
$H_2^+(2)$	Ar(40)	200	1,1°	P1	A	26	Tab.1	389
$H_2^+(2)$	Ar(40)	200-1200	1,1°	P1	A	29	Fig.4	25
$H_2^+(2)$	Ar(40)	500-1200	1,1°	P1	A	29	Tab.	26

TABLE 4.2

1	2	3	4	5	6	7	8	9
$H_2^+(2)$	Ar(40)	3	1	P2,3e	R	43	Fig.3	
$H_2^+(2)$	Ar(40)	10	1	P2,3e	R	42	Fig.2	635
$H_2^+(2)$	Ar(40)	10	1	P2,3e	R	43	Fig.2	
$H_2^+(2)$	Ar(40)	24	2	P2	R	24	Fig.6	
$H_2^+(2)$	Ar(40)	24	1,1⁻	P2	R	24	Fig.6	816
$H_2^+(2)$	Ar(40)	10	1	P3	R	104	Fig.4	
$H_2^+(2)$	Ar(40)	10	1	P3e	R	43	Fig.4	352
$H_2^+(2)$	Ar(40)	10.2	1	P3	R	44	Fig.11	346
$H_2^+(2)$	Ar(40)	20.4	1	P3	R	44	Fig.3	350-352
$H_2^+(2)$	Ar(40)	20.4	1	P3	R	44	Fig.7,8,9	349
$H_2^+(2)$	Ar(40)	20.4	1	P3	R	44	Fig.5	481
$H_2^+(2)$	Ar(40)	70	1	P3	R	92	Fig.6	
$H_2^+(2)$	Ar(40)	70	1	P3	R	105	Fig.17	
$H_2^+(2)$	Ar(40)	30-250	1,1°,2°	P4	A	48	Fig.1	
$H_2^+(2)$	Ar(40)	930	1°	P4	ND	106	Text	316
$H_2^+(2)$	Ar(40)	1000-3500	1°	P4	A	49	Fig.2	170
$H_2^+(2)$	Ar(40)	40-750	1°,2°	P4	A	46	Dwg.3	
$H_2^+(2)$	Ar(40)	40-2000	1	P4	A	46	Dwg.2	
$H_2^+(2)$	Ar(40)	40-2600	1,1°	P4	A	46	Dwg.1	
$H_2^+(2)$	Ar(40)	20,900	1,1°	P4	A	37	Tab.1	
$H_2^+(2)$	Ar(40)	9000-18000	none	P5	A	35	Tab.1	
$H_2^+(2)$	Ar(40)	10000-65000	none	P5	A	37	Fig.6	
$H_2^+(2)$	Ar(40)	200-1200	1,1°	P6	A	29	Fig.4	25

TABLE 4.2

1	2	3	4	5	6	7	8	9
$H_2^+(2)$	Kr(84)	2-50	1	P1	ND	25	Fig.5	68
$H_2^+(2)$	Kr(84)	2-50	$1,1^-$	P1	A	91	Tab.13	168
$H_2^+(2)$	Kr(84)	200	$1,1°$	P1	A	26	Tab.1	389
$H_2^+(2)$	Kr(84)	200-1200	$1,1°$	P1	A	29	Fig.4	25
$H_2^+(2)$	Kr(84)	500-1200	$1,1°$	P1	A	29	Tab.	26
$H_2^+(2)$	Kr(84)	10	1	P2,3e	R	42	Fig.2	635
$H_2^+(2)$	Kr(84)	10	1	P3e	R	43	Fig.4	
$H_2^+(2)$	Kr(84)	200-1200	$1,1°$	P6	A	29	Fig.4	25
$H_2^+(2)$	Xe(131)	2-50	1	P1	ND	25	Fig.5	68
$H_2^+(2)$	Xe(131)	2-50	$1,1^-$	P1	A	91	Tab.13	168
$H_2^+(2)$	Xe(131)	200	$1,1°$	P1	A	26	Tab.1	389
$H_2^+(2)$	Xe(131)	10	1	P3c	R	107	Fig.2	
$H_2^+(2)$	Xe(131)	10	1	P3c	R	107	Fig.3	
$H_2^+(2)$	Xe(131)	10	1	P3	R	104	Fig.4	816
$H_2^+(2)$	Xe(131)	14	1	P3	R	38	Fig.3	
$H_2^+(2)$	Xe(131)	20.4	1	P3	R	44	Fig.7,8,9	350-352
$H_2^+(2)$	$H_2(2)$	2.0	1	P1a	R	10	Fig.1a	974
$H_2^+(2)$	$H_2(2)$	2.0	1	P1a	R	10	Fig.1b	974
$H_2^+(2)$	$H_2(2)$	2-50	1^-	P1	ND	25	Fig.6	68
$H_2^+(2)$	$H_2(2)$	2-50	1	P1	ND	25	Fig.4	67
$H_2^+(2)$	$H_2(2)$	3.0	1	P1	R	108	Tab.	244
$H_2^+(2)$	$H_2(2)$	3-115	1	P1	ND	6	Tab.1	183
$H_2^+(2)$	$H_2(2)$	3.2-120	$1,1°,2°$	P1c	ND	27	Fig.6	1856
$H_2^+(2)$	$H_2(2)$	5-29	1	P1	A	23	Fig.11	

TABLE 4.2

1	2	3	4	5	6	7	8	9
$H_2^+(2)$	$H_2(2)$	5-50	1	Plc	A	109		
$H_2^+(2)$	$H_2(2)$	6,20,50	$1,1°,1^-$	Plc	ND	25	Fig.5	66
$H_2^+(2)$	$H_2(2)$	10-180	1	Pl	A	100	Fig.1	
$H_2^+(2)$	$H_2(2)$	15-90	$1,1°,2°$	Pl	A	69	Fig.7,8	557
$H_2^+(2)$	$H_2(2)$	30-240	1	Pld	A	28	Fig.4	
$H_2^+(2)$	$H_2(2)$	40-200	1	Pl	R	28	Fig.5	
$H_2^+(2)$	$H_2(2)$	50-200	1	Plb	R	28	Fig.6	
$H_2^+(2)$	$H_2(2)$	1.7	1	Pla	R	9	Fig.12	339
$H_2^+(2)$	$H_2(2)$	100-200	$1,1°$	Pl	A	26	Tab.3	390
$H_2^+(2)$	$H_2(2)$	200	$1,1°$	Pl	A	26	Tab.1	389
$H_2(2)$	$H_2(2)$	200-600	1	Pl	A	30	Fig.4	6
$H_2(2)$	$H_2(2)$	200-1200	$1,1°$	Pl	A	29	Fig.2	25
$H_2(2)$	$H_2(2)$	500-1200	$1,1°$	Pl	A	29	Tab.	26
$H_2(2)$	$H_2(2)$	?	1	Pl	A	7	Text	1084
$H_2(2)$	$H_2(2)$	?	1	Pla	R	7	Fig.1	1084
$H_2(2)$	$H_2(2)$	5	$1,1°$	P2	NS	63	Fig.1a	771
$H_2(2)$	$H_2(2)$	10	1	P2	R	40	Fig.4	263
$H_2^+(2)$	$H_2(2)$	10	1	P2	ND	6	Text	183
$H_2^+(2)$	$H_2(2)$	10	$1,1°$	P2	NS	63	Fig.1b	771
$H_2^+(2)$	$H_2(2)$	20	$1,1°$	P2	NS	63	Fig.1c	771
$H_2^+(2)$	$H_2(2)$	40	$1,1°$	P2	NS	63	Fig.1d	771
$H_2^+(2)$	$H_2(2)$	40-200	1	P2	A	28	Fig.3	
$H_2^+(2)$	$H_2(2)$	80	$1,1°$	P2	NS	63	Fig.1e	771
$H_2^+(2)$	$H_2(2)$	200	$1°$	P2	R	26	Fig.6	392

TABLE 4.2

1	2	3	4	5	6	7	8	9
$H_2^+(2)$	$H_2(2)$	200	1°,2°	P2	R	21	Fig.4	420
$H_2^+(2)$	$H_2(2)$	200	1	P2	R	30	Fig.6	7
$H_2^+(2)$	$H_2(2)$	200-600	1	P2	R	30	Fig.7	8
$H_2^+(2)$	$H_2(2)$	0.75-2.0	1	P3	R	110	Fig.1	626
$H_2^+(2)$	$H_2(2)$	2	1	P3	R	9	Fig.9	332
$H_2^+(2)$	$H_2(2)$	3-20	1	P3,4	R	111	Fig.4,8	22-24
$H_2^+(2)$	$H_2(2)$	5-23	1	P3	R	112	Fig.2	177
$H_2^+(2)$	$H_2(2)$	10	1	P3	R	40	Fig.7abc	268
$H_2^+(2)$	$H_2(2)$	10	1	P3e	R	43	Fig.2	
$H_2^+(2)$	$H_2(2)$	10	1	P3	R	40	Fig.3	262
$H_2^+(2)$	$H_2(2)$	10	1	P3c	R	107	Fig.2	
$H_2^+(2)$	$H_2(2)$	10	1	P3c	R	107	Fig.3	
$H_2^+(2)$	$H_2(2)$	10	1	P3d	R	104	Fig.4,6,7	816
$H_2^+(2)$	$H_2(2)$	14	1	P3	R	38	Fig.3	
$H_2^+(2)$	$H_2(2)$	20.4	1	P3	R	44	Fig.10	352
$H_2^+(2)$	$H_2(2)$	70	1	P3	R	92	Fig.2	481
$H_2^+(2)$	$H_2(2)$	70	1	P3	R	105	Fig.17	
$H_2^+(2)$	$H_2(2)$	30-250	1,1°,2°	P4	A	48	Fig.1	
$H_2^+(2)$	$H_2(2)$	30-780	1°,2°	P4	A	46	Dwg.3	
$H_2^+(2)$	$H_2(2)$	30-3000	1,1°	P4	A	46	Dwg.1	
$H_2^+(2)$	$H_2(2)$	30-3000	1	P4	A	46	Dwg.2	
$H_2^+(2)$	$H_2(2)$	100-200	1°,2°	P4	A	21	Fig.7	423
$H_2^+(2)$	$H_2(2)$	100-800	1,1°	P4	A	21	Fig.5	422
$H_2^+(2)$	$H_2(2)$	100-800	1,1°	P4	A	21	Fig.6	422

TABLE 4.2

1	2	3	4	5	6	7	8	9
$H_2^+(2)$	$H_2(2)$	280-670	1,1°	P4b	A	47	Fig.1	1216
$H_2^+(2)$	$H_2(2)$	3000	1°	P4	A	113	Tab.1	233
$H_2^+(2)$	$H_2(2)$	20,900	1,1°	P4	A	37	Tab.1	188
$H_2^+(2)$	$H_2(2)$	10,15,20		P5	A	114	Fig.1	
$H_2^+(2)$	$H_2(2)$	40-220	1,2	P5	A	115	Fig.6.2	81
$H_2^+(2)$	$H_2(2)$	200-600	none	P5	A	30	Fig.3	5
$H_2^+(2)$	$H_2(2)$	10000-65000	none	P5	A	37	Fig.3	
$H_2^+(2)$	$H_2(2)$	15-90	1,1°,2°	P6	A	69	Fig.7,8	557
$H_2^+(2)$	$H_2(2)$	200-1200	1,1°	P6	A	29	Fig.2	25
$H_2^+(2)$	$D_2(4)$	3.0	1	P1	R	108	Tab.	244
$H_2^+(2)$	$D_2(4)$	200	1,1°	P1	A	26	Tab.1	389
$H_2^+(2)$	$N_2(28)$	2-50	1	P1	ND	25	Fig.5	68
$H_2^+(2)$	$N_2(28)$	5-29	1	P1	A	23	Fig.11	
$H_2^+(2)$	$N_2(28)$	10-180	1	P1	A	100	Fig.2	
$H_2^+(2)$	$N_2(28)$	150,200	1,1°	P1	A	26	Tab.3	390
$H_2^+(2)$	$N_2(28)$	200	1,1°	P1	A	26	Tab.1	389
$H_2^+(2)$	$N_2(28)$	200-1200	1,1°	P1	A	29	Fig.3	25
$H_2^+(2)$	$N_2(28)$	500-1200	1,1°	P1	A	29	Tab.	26
$H_2^+(2)$	$N_2(28)$	30-250	1,1°,2°	P4	A	48	Fig.1	
$H_2^+(2)$	$N_2(28)$	35-3000	1,1°	P4	A	46	Dwg.1	
$H_2^+(2)$	$N_2(28)$	35-3000	1	P4	A	46	Dwg.2	
$H_2^+(2)$	$N_2(28)$	36-1000	1°,2°	P4	A	46	Dwg.3	
$H_2^+(2)$	$N_2(28)$	3000	1°	P4	A	113	Tab.1	233
$H_2^+(2)$	$N_2(28)$	20,900	1,1°	P4	A	37	Tab.1	

TABLE 4.2

1	2	3	4	5	6	7	8	9
$H_2^+(2)$	$N_2(28)$	10000-65000	none	P5	A	37	Fig.5	
$H_2^+(2)$	$N_2(28)$	200-1200	1,1°	P6	A	29	Fig.3	25
$H_2^+(2)$	Air(~30)	5-180	1	P1	A	24	Fig.5	244
$H_2^+(2)$	Air(~30)	30	1	P1	R	108	Tab.	391
$H_2^+(2)$	Air(~30)	100	1	P1	A	18	text	
$H_2^+(2)$	Air(~30)	9000-18000	none	P5	A	35	Tab.1	389
$H_2^+(2)$	$O_2(32)$	200	1,1°	P1	A	26	Tab.1	
$H_2^+(2)$	$O_2(32)$	50-1000	1°,2°	P4	A	46	Dwg.3	390
$H_2^+(2)$	$O_2(32)$	50-3000	1,1°	P4	A	46	Dwg.1	390
$H_2^+(2)$	$O_2(32)$	50-3000	1	P4	A	46	Dwg.2	390
$H_2^+(2)$	$H_2S(34)$	200	1,1°	P1	A	26	Tab.2	390
$H_2^+(2)$	$CO_2(44)$	200	1,1°	P1	A	26	Tab.2	390
$H_2^+(2)$	$N_2O(44)$	200	1,1°	P1	A	26	Tab.2	390
$H_2^+(2)$	$CH_4(16)$	200	1,1°	P1	A	26	Tab.2	390
$HD^+(3)$	He(4)	.004-.048	1,2	P1	A	116	Tab.1	2691
$HD^+(3)$	He(4)	4.25	1,2	P1	A	117	text	331
$HD^+(3)$	He(4)	4.25	1,2	P3	R	117	text	330
$HD^+(3)$	Ne(20)	.004-.048	1,2	P1	A	116	Tab.1	2691
$HD^+(3)$	Ne(20)	4.25	1,2	P1	A	117	text	331
$HD^+(3)$	Ar(40)	.004-.048	1,2	P1	A	116	Tab.1	2691
$HD^+(3)$	Ar(40)	4.25	1,2	P1	A	117	text	331
$HD^+(3)$	Kr(84)	.004-.048	1,2	P1	A	116	Tab.1	2691
$HD^+(3)$	Xe(131)	.004-.048	1,2	P1	A	116	Tab.1	2691
$HD^+(3)$	$H_2(2)$	3.0	1,2	P1	R	108	Tab.	244

TABLE 4.2

1	2	3	4	5	6	7	8	9
$HD^+(3)$	$D_2(4)$	3.0	1,2	P1	R	108	Tab.	244
$HD^+(3)$	Air(~30)	3.0	1,2	P1	R	108	Tab.	244
$D_2^+(4)$	Ar(40)	0.051	2	P2	R	79	Fig.7b	
$D_2^+(4)$	Ar(40)	0.01-0.10	2	P3	R	79	Fig.6	
$D_2^+(4)$	Ar(40)	0.045	2	P3	R	79	Fig.4b	
$D_2^+(4)$	Ar(40)	0.075	2	P3	R	79	Fig.4a	
$D_2^+(4)$	Ar(40)	20.4	2	P3	R	44	Fig.11	352
$D_2^+(4)$	$H_2(2)$	3.0	2	P1	R	108	Tab.	244
$D_2^+(4)$	$H_2(2)$	20-90	2,2°,4°	P1	A	69	Fig.7,9	557
$D_2^+(4)$	$H_2(2)$	24-90	2,2°,4°	P1	A	69	Fig.7,9	557
$D_2^+(4)$	$H_2(2)$	14	2	P3	R	38	Fig.4	244
$D_2^+(4)$	$D_2(4)$	3.0	2	P1	R	108	Tab.	244
$D_2^+(4)$	$D_2(4)$	3.5-100	2	P1	A	118	Tab.	1152
$D_2^+(4)$	$D_2(4)$	.001-.050	2	P3	R	119	Fig.7,8	392
$D_2^+(4)$	$D_2(4)$	10-100	2	P1	A	64	Fig.4	934
$D_2^+(4)$	$D_2(4)$	10-100	2	P3	R	64	Fig.6	935
$D_2^+(4)$	$N_2(28)$	0.02-0.10	2	P2	none	120		
$D_2^+(4)$	$N_2(28)$	0.065	2	P2	R	79	Fig.7a	
$D_2^+(4)$	$N_2(28)$	0.01-0.08	2	P3	R	79	Fig.6	
$D_2^+(4)$	$N_2(28)$	0.02-0.10	2	P3	none	120		
$D_2^+(4)$	$N_2(28)$	0.026	2	P3	R	79	Fig.3b	
$D_2^+(4)$	$N_2(28)$	0.065	2	P3	R	79	Fig.3a	
$D_2^+(4)$	Air(~30)	3.0	2	P1[1]	R	108	Tab.	244
$D_2^+(4)$	Air(~30)	4.4	2	P1a	R	62	Fig.3	1860

TABLE 4.2

1	2	3	4	5	6	7	8	9
$D_2^+(4)$	$O_2(32)$	0.02-0.10	2	P2	none	120		
$D_2^+(4)$	$O_2(32)$	0.02-0.10	2	P3	none	120		
$HeH^+(5)$	He(4)	2.4	1	P2,3	R	121	Fig.3	
$HeH^+(5)$	He(4)	3.0	4	P3	R	121	Fig.4	
$HeH^+(5)$	He(4)	3.5	1	P3	R	121	Fig.1	
$HeH^+(5)$	He(4)	10	4	P3	R	122	Fig.1	
$HeH^+(5)$	Air(~28)	5000	$1°,4°$	P5	NS	36	Tab.	402
$HeH^+(5)$	Ar(40)	10	1	P3	R	123	Fig.1	
$HeH^+(5)$	Kr(84)	10	4	P3	R	122	Fig.1	
$HeH^+(5)$	Xe(131)	3.0	4	P3	R	121	Fig.4	
$HeH^+(5)$	?	10	1	P2,3	R	124	Fig.3-12	
$CH^+(13)$	Ne(20)	3.5	12	P1a	A	125	Text,Fig.	1499
$CH^+(13)$	Air(~30)	2.5	12	P1	R	8	Tab.	460
$CH^+(13)$	Air(~30)	3.5,90	12	P1	R	62	Fig.2	1859
$CH^+(13)$	Air(~30)	3.5	12	P1	A	126	Tab.3	1077
$NH^+(15)$	Air(~30)	3.5	14	P1	A	126	Tab.3	1077
$OH^+(17)$	Air(~30)	3.5	16	P1	A	126	Tab.3	1077
$C_2^+(24)$	Ne(20)	3.5	12	P1	A	127	Tab.1	1019
$C_2^+(24)$	Ne(20)	3.5	12	P1	A	126	Tab.3	1077
$C_2^+(24)$	Air(~30)	3.5	12	P1	A	126	Tab.3	1077
$CO^+(28)$	He(4)	2.1,4.1,5.1	12	P1	R	99	Tab.3	1136
$CO^+(28)$	He(4)	5.1,2.1	16	P1	R	99	Tab.4	1137
$CO^+(28)$	He(4)	15	16	P3	R	31	Fig.3	196
$CO^+(28)$	He(4)	17.5-30	16	P3	R	31	Tab.4	198

42

TABLE 4.2

1	2	3	4	5	6	7	8	9
CO^+(28)	Ne(20)	1.65-2.7	12,16	P1b	R	15	Tab.1,2	207-208
CO^+(28)	Ar(40)	2.1,4.1,5.1	12	P1	R	99	Tab.3	1136
CO^+(28)	Ar(40)	5.1,2.1	16	P1	R	99	Tab.4	1137
CO^+(28)	Kr(84)	2.1,4.1,5.1	12	P1	R	99	Tab.3	1136
CO^+(28)	Kr(84)	5.1,2.1	16	P1	R	99	Tab.4	1137
CO^+(28)	H_2(2)	2.1,4.1,5.1	12	P1	R	99	Tab.3	1136
CO^+(28)	H_2(2)	5.1,2.1	16	P1	R	99	Tab.4	1137
CO^+(28)	D_2(4)	2.1,4.1,5.1	12	P1	R	99	Tab.3	1136
CO^+(28)	N_2(28)	2.1,4.1,5.1	12	P1	R	99	Tab.3	1136
CO^+(28)	CO(28)	2.1,4.1,5.1	12	P1	R	99	Tab.3	1136
CO^+(28)	CO(28)	5.1,2.1	16	P1	R	99	Tab.4	1137
CO^+(28)	Air(~30)	2.5	12,16	P1	R	8	Tab.	460
CO^+(28)	CO_2(44)	1.0	12	P1	ND	17	text	2270
						93	text	2,51
N_2^+(28)	He(4)	17.5-30	14	P3	R	31	Tab.4	198
N_2^+(28)	Ne(20)	5-25	14	P1	A	23	Fig.11	
N_2^+(28)	Ar(40)	5-25	14	P1	A	23	Fig.11	
N_2^+(28)	H_2(2)	5-25	14	P1	A	23	Fig.11	2089
N_2^+(28)	N_2(28)	0.6-2.0	14	P1	A	11	Fig.3	2270
N_2^+(28)	N_2(28)	1.0	14	P1	ND	17	text	2,56
						93	text	
N_2^+(28)	N_2(28)	1.6	14	P1a	R	11	Fig.6	2094
N_2^+(28)	N_2(28)	1.6	14	P1a	R	11	Tab.3	2094
N_2^+(28)	N_2(28)	5-25	14	P1	A	23	Fig.11	

TABLE 4.2

1	2	3	4	5	6	7	8	9
$N_2^+(28)$	Air(28)	2.8	14	P1a		128	Fig.1	1343
$N_2^+(28)$	$N_2(28)$	2.0	14	P3	R	9	Fig.4	325
$N_2^+(29\ )$	$N_2(28)$	2.0	14,15	P3	R	9	Fig.5	326
$NO^+(30)$	He(4)	17.5-30	16	P3	R	31	Tab.4	198
$O_2^+(32)$	He(4)	.03-.35	16	P1	A	98	Fig.12	93
$O_2^+(32)$	He(4)	.040-.36	16	P1	A	129	Fig.1	2544
$O_2^+(32)$	He(4)	15	$16,16^{+2},16^{-}$ $32^{+2},32^{-}$	P1	A	31	Tab.3	195
$O_2^+(32)$	Ar(40)	15	$16,16^{+2},16^{-}$ 32^{-}	P1	A	31	Tab.3	195
$O_2^+(32)$	$H_2(2)$	.03-.35	16	P1	A	98	Fig.14	96
$O_2^+(32)$	$H_2(2)$	.040-.36	16	P1	A	129	Fig.1	2544
$O_2^+(32)$	Air(28)	2.8	$16,16^{-}$	P1a		128	Fig.1	1343
$O_2^+(32)$	$O_2(32)$	0.6-1.8	16	P1	A	11	Fig.3	2089
$O_2^+(32)$	$O_2(32)$	2.0	16	P1a	R	9	Fig.11	338
$O_2^+(32)$	$O_2(32)$	1.6	16	P1a	R	11	Tab.1	2090
$O_2^+(32)$	$O_2(32)$	1.6	16	P1a	R	11	Fig.1	2087
$O_2^+(32)$	$O_2(32)$	15	$16^{+},16^{+2},16^{-}$ 32^{+}	P1	A	31	Tab.3	195
$O_2^+(32)$	$O_2(32)$	1.6	16	P3	R	9	Fig.6a	329
$SO^+(48)$	He(4)	17.5-30	16	P3	R	31	Tab.4	198
$SF^+(51)$	He(4)	17.5-30	19	P3	R	31	Tab.4	198

TABLE 4.2

1	2	3	4	5	6	7	8	9
n=2	q=0							
$H_2(2)$	$He^{+2}(4)$	3200	1,2,3	T1	R	95	Tab.	81
$H_2(2)$	$He^{+2}(4)$	3200	1,2,3	T1	R	95	Fig.12	48
$H_2(2)$	$Kr^{+2}(84)$	.025-.90	2,1	T1	ND	130	Tab.2,3	225,227
$H_2(2)$	$Ar^{+2}(40)$	.025-.90	2,1	T1	ND	130	Tab.2,3	225,227
$H_2(2)$	$H^+(1)$	.025-.90	2,1	T1	ND	130	Tab.2,3	225,227
$H_2(2)$	$H^+(1)$	5-45	1,2	T1	A	78	Fig.4	1153
$H_2(2)$	$H^+(1)$	5-180	2	T1	A	131	Fig.4	971
$H_2(2)$	$H^+(1)$	5-180	1	T1	A	131	Fig.5	971
$H_2(2)$	$H^+(1)$	10-180	1,2	T1	A	132	Fig.5a	461
$H_2(2)$	$H^+(1)$	0.15-6	None	T4	ND	133	Fig.1	197
$H_2(2)$	$H^+(1)$	2-50	None	T4	A	88	Fig.2	8
$H_2(2)$	$H^+(1)$	4-50	None	T4	ND	85	Fig.5	391
$H_2(2)$	$H^+(1)$	6-49	None	T4	A	87	Fig.1	1636
$H_2(2)$	$H^+(1)$	400-1000	None	T4		86	Fig.4	
$H_2(2)$	$He^+(4)$	.025-.90	2,1	T1	ND	130	Tab.2,3	225,227
$H_2(2)$	$He^+(4)$	5.45	1,2	T1	ND	134	Fig.1	535
$H_2(2)$	$He^+(4)$	20-180	1,2	T1	A	68	Fig.4c	697
$H_2(2)$	$C^+(12)$	.025-.90	2,1	T1	ND	130	Tab.2,3	225,227
$H_2(2)$	$N^+(14)$	.025-.90	2,1	T1	ND	130	Tab.2,3	225,227
$H_2(2)$	$O^+(16)$	.025-.90	2,1	T1b	ND	130	Tab.2,3	225,227
$H_2(2)$	$F^+(19)$	.025-.90	2,1	T1	ND	130	Tab.2,3	225,227
$H_2(2)$	$Ne^+(20)$	.025-.90	2,1	T1	ND	130	Tab.2,3	225,227

TABLE 4.2

1	2	3	4	5	6	7	8	9
H$_2$(2)	S$^+$(32)	.025-.90	2,1	T1	ND	130	Tab.2,3	225,227
H$_2$(2)	Cl$^+$(35)	.025-.90	2,1	T1	ND	130	Tab.2,3	225,227
H$_2$(2)	Ar$^+$(40)	.025-.90	2,1	T1	ND	130	Tab.2,3	225,227
H$_2$(2)	Kr$^+$(84)	.025-.90	2,1	T1	ND	130	Tab.2,3	225,227
H$_2$(2)	Kr$^+$(84)	0.5	1	T1	ND	72	Tab.1	1214
H$_2$(2)	H(1)	10-180	1,2	T1	A	132	Fig.5a	461
H$_2$(2)	He(4)	20-180	1,2	T1	A	68	Fig.4c	697
H$_2$(2)	H$_2^+$(2)	.01-.1	1	T1 / P1	A	45	Fig.7	490
H$_2$(2)	H$_2^+$(2)	.025-.90	2,1	T1	ND	130	Tab.2,3	225,227
H$_2$(2)	H$_2^+$(2)	0.5-2.0	1,2	T1	ND	72	Tab.1	121
H$_2$(2)	H$_2^+$(2)	1.4-46	1,2	T1a	A	135	Fig.2	1057
H$_2$(2)	H$_2^+$(2)	5-180	1	T1	A	131	Fig.5	971
H$_2$(2)	H$_2^+$(2)	5-180	2	T1	A	131	Fig.4	971
H$_2$(2)	N$_2^+$(28)	.05-.1	1	T1	A	45	Fig.8	490
H$_2$(2)	O$_2^+$(32)	15	None	T4	A	31	Tab.3	195
H$_2$(2)	H$_2$(2)	6-120	1,1°,2	P1	ND	65	Fig.4	1232
H$_2$(2)	H$_3^+$(3)	5-180	2	T1	A	131	Fig.4	971
H$_2$(2)	H$_3^+$(3)	5-180	1	T1	A	131	Fig.5	971
CO(28)	e	0.5	12,16,28	T1	ND	136	Tab.	434
CO(28)	Ne^{+3}(20)	20-90	12,16,28	T1	A	67	Fig.5	87
CO(28)	He^{+2}(4)	3200	12,14,16, 28,30	T1	R	95	Tab.	76
CO(28)	Ne^{+2}(20)	20-60	16	T1	A	67	Fig.8	88

TABLE 4.2

1	2	3	4	5	6	7	8	9
CO(28)	$Ar^{+2}(40)$	.025-.90	28,16,12	T1	ND	130	Tab.6,7,8	234,235
CO(28)	$Ar^{+2}(40)$	0.05	12,16,28	T1	ND	136	Tab.	434
CO(28)	$Kr^{+2}(84)$	.025-.90	28,12	T1	ND	130	Tab.6,7,8	234,235
CO(28)	$Kr^{+2}(84)$	0.5	12,16,28	T1	ND	136	Tab.	434
CO(28)	$H^{+}(1)$	.025-.90	28,16,12	T1	ND	130	Tab.6,7,8	234,235
CO(28)	$H^{+}(1)$	5-45	12,16,28, $12^{+2},16^{+2}$ 28^{+2}	T1	ND	78	Fig.7	1154
CO(28)	$He^{+}(4)$	.001-.037	12	T1	A	137	Fig.1	
CO(28)	$He^{+}(4)$	.025-.90	28,16,12	T1	ND	130	Tab.6,7,8	234,235
CO(28)	$He^{+}(4)$	0.5	12,16,28	T1	ND	136	Tab.	434
CO(28)	$He^{+}(4)$	5-45	12,16,28, $12^{+2},16^{+2}$, 28^{+2}	T1	ND	134	Fig.4	539
CO(28)	$B^{+}(11)$	.025-.90	28,16,12	T1	ND	130	Tab.6,7	234,235
CO(28)	$B^{+}(11)$	0.5	12,16,28	T1	ND	136	Tab.	434
CO(28)	$C^{+}(12)$	.025-.90	28,16,12	T1	ND	130	Tab.6,7,8	234,235
CO(28)	$C^{+}(12)$	0.5	12,16,28	T1	ND	136	Tab.	434
CO(28)	$N^{+}(14)$	.025-.90	28,16,12	T1	ND	130	Tab.6,7,8	234,235
CO(28)	$N^{+}(14)$	0.5	12,16,28	T1	ND	136	Tab.	434
CO(28)	$F^{+}(19)$	.025-.90	28,16,12	T1	ND	130	Tab.6,7,8	234,235
CO(28)	$F^{+}(19)$	0.5	12,16,28	T1	ND	136	Tab.	434
CO(28)	$Ne^{+}(20)$	.001-040	12	T1	A	137	Fig.2	
CO(28)	$Ne^{+}(20)$	.025-.90	28,16,12	T1	ND	130	Tab.6,7,8	234,235

TABLE 4.2

1	2	3	4	5	6	7	8	9
CO(28)	$Ne^+(20)$	0.5	12,16,28	T1	ND	136	Tab.	434
CO(28)	$Ne^+(20)$	1-30	12,16,28	T1	A	67	Fig.3	85
CO(28)	$Ne^+(20)$	6-30	$12^-,16^-28^{-2}$	T1	A	67	Fig.6	87
CO(28)	$P^+(31)$	0.5	12,16,28	T1	ND	136	Tab.	434
CO(28)	$S^+(32)$	0.5	12,16,28	T1	ND	136	Tab.	434
CO(28)	$Cl^+(35)$	.025-.90	28,16,12	T1	ND	130	Tab.6,7,8	234,235
CO(28)	$Cl^+(35)$	0.5	12,16,28	T1	ND	136	Tab.	434
CO(28)	$Ar^+(40)$	.015-.063	12	T1	A	137	Fig.3	2179
CO(28)	$Ar^+(40)$	.016-019	12	T1	A	138	Fig.5	2179
CO(28)	$Ar^+(40)$	.023-093	12	T1	ND	77	Fig.2	1153
CO(28)	$Ar^+(40)$	.025-.90	28,12	T1	ND	130	Tab.6,7,8	234,235
CO(28)	$Ar^+(40)$	0.5	12,16,28	T1	ND	136	Tab.	434
CO(28)	$Br^+(80)$	0.5	12,16,28	T1	ND	136	Tab.	434
CO(28)	$Kr^+(84)$	0.5	12,16,28	T1	ND	136	Tab.	434
CO(28)	H(1)	5-28	16^{-2}	T1	R	66	Fig.5	270
CO(28)	H(1)	5-30	12,16,28	T1	R	66	Fig.9	272
CO(28)	Ne(20)	4-30	12,16,28	T1	A	67	Fig.2	85
CO(28)	Ne(20)	10-30	$12^-,16^{-2}$	T1	A	67	Fig.7	87
CO(28)	$H_2^+(2)$	.025-.90	28,16,12	T1	ND	130	Tab.6,7,8	234,235
CO(28)	$CO^+(28)$	0.5	12,16,28	T1	ND	136	Tab.	434
$N_2(28)$	e	0.5	14,28	T1	ND	139	Tab.	258
$N_2(28)$	$S^{+2}(32)$	0.5	14,28	T1	ND	139	Tab.	258
$N_2(28)$	$Ar^{+2}(40)$	.025-.90	28,14	T1	ND	130	Tab.4,5	230,231
$N_2(28)$	$Ar^{+2}(40)$	0.5	14,28	T1	ND	139	Tab.	258
$N_2(28)$	$Kr^{+2}(84)$	.025-.90	28,14	T1	ND	130	Tab.4,5	230,231

TABLE 4.2

1	2	3	4	5	6	7	8	9
$N_2(28)$	$Kr^{+2}(84)$	0.5	14,28	T1	ND	139	Tab.	258
$N_2(28)$	$H^+(1)$	.025-.90	28,14	T1	ND	130	Tab.4,5	230,231
$N_2(28)$	$H^+(1)$	3-37	14	T1	R	140	Fig.2	1127
$N_2(28)$	$H^+(1)$	5-45	$7,14,7^{+2}$	T1	ND	78	Fig.6	1153
$N_2(28)$	$H^+(1)$	5-180	$14,14^{+2},28$	T1	A	141	Fig.8	32
$N_2(28)$	$H^+(1)$	10-180	$14,14^{+2},28$	T1	A	132	Fig.5b	461
$N_2(28)$	$He^+(4)$	.0005-.04	14	T1	A	142	Fig.2	
$N_2(28)$	$He^+(4)$	.001-.04	28	T1	A	142	Fig.3	
$N_2(28)$	$He^+(4)$	.003-0.1	14,28	T1	AN	75	Tab.1	1127
$N_2(28)$	$He^+(4)$	.025-.90	28,14	T1	ND	130	Tab.4,5	230,231
$N_2(28)$	$He^+(4)$	0.4-7.6	14,28	T1	A	76	Fig.3	970
$N_2(28)$	$He^+(4)$	0.5	14,28	T1	ND	139	Tab.	258
$N_2(28)$	$He^+(4)$	0.5-2.0	14	T1	ND	94	Tab.V	37
$N_2(28)$	$He^+(4)$	0.5-2.0	28	T1	ND	94	Tab.IV	36
$N_2(28)$	$He^+(4)$	0.5-2.0	14,28	T1	ND	72	Tab.1	1214
$N_2(28)$	$He^+(4)$	3-37	14	T1	R	140	Fig.2	1127
$N_2(28)$	$He^+(4)$	5-45	$14,28,14^{+2}$	T1	ND	134	Fig.2	537
$N_2(28)$	$He^+(4)$	15-180	$14,14^{+2},28$	T1	A	68	Fig.4d	697
$N_2(28)$	$B^+(11)$	0.5	14,28	T1	ND	139	Tab.	258
$N_2(28)$	$C^+(12)$	.025-.90	14,28	T1	ND	130	Tab.4,5	230,231
$N_2(28)$	$C^+(12)$	0.5	14,28	T1	ND	139	Tab.	258
$N_2(28)$	$N^+(14)$	.025-.90	14,28	T1	ND	130	Tab.4,5	230,231
$N_2(28)$	$N^+(14)$	0.5	14,28	T1	ND	139	Tab.	258
$N_2(28)$	$O^+(16)$	.025-.90	14,28	T1	ND	130	Tab.4,5	230,231
$N_2(28)$	$O^+(16)$	0.5	14,28	T1b	ND	139	Tab.	258

49

TABLE 4.2

1	2	3	4	5	6	7	8	9
$N_2(28)$	$F^+(19)$	.025-.90	14,28	T1	ND	130	Tab.4,5	230,231
$N_2(28)$	$F^+(19)$	0.5	14,28	T1	ND	139	Tab.	258
$N_2(28)$	$Ne^+(20)$	.005-.008	14	T1	A	138	Fig.6	2179
$N_2(28)$	$Ne^+(20)$	.005-.027	14	T1	A	138	Fig.1	2175
$N_2(28)$	$Ne^+(20)$	.025-.90	14,28	T1	ND	130	Tab.4,5	230,231
$N_2(28)$	$Ne^+(20)$	0.5	14,28	T1	ND	139	Tab.	258
$N_2(28)$	$Ne^+(20)$	0.5-2.0	14	T1	ND	94	Tab.V	37
$N_2(28)$	$Ne^+(20)$	0.5-2.0	28	T1	ND	94	Tab.IV	36
$N_2(28)$	$Ne^+(20)$	0.5-2.0	14,28	T1	ND	72	Tab.1	1214
$N_2(28)$	$Ne^+(20)$	3-37	14	T1	R	140	Fig.2	1127
$N_2(28)$	$S^+(32)$	.025-.90	14,28	T1	ND	130	Tab.4,5	230,231
$N_2(28)$	$S^+(32)$	0.5	14,28	T1	ND	139	Tab.	258
$N_2(28)$	$Cl^+(35)$	0.5	14,28	T1	ND	139	Tab.	258
$N_2(28)$	$Ar^+(40)$	.02-.05	14	T1	A	138	Fig.2	2176
$N_2(28)$	$Ar^+(40)$	.023-.093	14	T1	ND	77	Fig.2	1153
$N_2(28)$	$Ar^+(40)$	.025-.90	14,28	T1	ND	130	Tab.4,5	230,231
$N_2(28)$	$Ar^+(40)$	0.5	14,28	T1	ND	139	Tab.	258
$N_2(28)$	$Ar^+(40)$	3-37	14	T1	R	140	Fig.2	1127
$N_2(28)$	$Se^+(79)$	0.5	14,28	T1	ND	139	Tab.	258
$N_2(28)$	$Br^+(80)$	0.5	14,28	T1	ND	139	Tab.	258
$N_2(28)$	$Kr^+(84)$	.023-.093	14	T1	ND	77	Fig.2	1153
$N_2(28)$	$Kr^+(84)$	.025-.90	14,28	T1	ND	130	Tab.4,5	230,231
$N_2(28)$	$Kr^+(84)$	0.5	14,28	T1	ND	139	Tab.	258
$N_2(28)$	$Xe^+(131)$	.023-.093	14	T1	ND	77	Fig.2	1153
$N_2(28)$	$H(1)$	3-37	14	T1	R	140	Fig.2	1127

TABLE 4.2

1	2	3	4	5	6	7	8	9
$N_2(28)$	$H(1)$	7-28	$14,28^3$	T1	R	66	Fig.6	271
$N_2(28)$	$H(1)$	10-180	$14,14^{+2},28$	T1	A	132	Fig.5b	461
$N_2(28)$	$He(4)$	3-37	14	T1	R	140	Fig.2	1127
$N_2(28)$	$He(4)$	15-180	$14,14^{+2},28$	T1	A	68	Fig.4d	697
$N_2(28)$	$Ne(20)$	3-37	14	T1	R	140	Fig.2	1127
$N_2(28)$	$Ar(40)$	3-37	14	T1	R	140	Fig.2	1127
$N_2(28)$	$H_2^+(2)$	.025-.90	28,14	T1	ND	130	Tab.4,5	230,231
$N_2(28)$	$H_2^+(1)$	5-180	$14,14^{+2},28$	T1	A	141	Fig.8	32
$N_2(28)$	$N_2^+(28)$	.015-043	14	T1	A	143	Fig.1	859
$N_2(28)$	$N_2^+(28)$	.025-.90	14,28	T1	ND	130	Tab.4,5	230,231
$N_2(28)$	$N_2^+(28)$	0.5	14,28	T1	ND	139	Tab.	258
$NO(30)$	$Ar^+(40)$	.023-093	14,16	T1	ND	77	Fig.4	1153
$NO(30)$	$Kr^+(84)$	.023-093	14,16	T1	ND	77	Fig.4	1153
$NO(30)$	$Xe^+(131)$	.023-093	14	T1	ND	77	Fig.4	1153
$NO(30)$	$H(1)$	4-32	$16^-,30^{-2}$	T1	R	66	Fig.4	269
$NO(30)$	$H(1)$	5-30	$14,16,30^3$	T1	R	66	Fig.8	272
$O_2(32)$	$H^+(1)$	5-45	$16,32,16^{+2}$	T1	ND	78	Fig.5	1153
$O_2(32)$	$H^+(1)$	5-180	$16,16^{+2},32$	T1	A	141	Fig.9	32
$O_2(32)$	$He^+(4)$	.0005-.4	16	T1	A	142	Fig.1	1128
$O_2(32)$	$He^+(4)$	.007-0.1	16	T1	AN	75	Fig.3	279
$O_2(32)$	$He^+(4)$	.018	16,32	T1	R	144	Tab.2	279
$O_2(32)$	$He^+(4)$	0.1-5.6	16,32	T1	A	76	Fig.2	970
$O_2(32)$	$He^+(4)$	5-45	$16,32,16^{+2}$	T1	ND	134	Fig.3	537
$O_2(32)$	$F^+(19)$	.025-.9	16,32	T1	R	144	Tab.2	279

51

TABLE 4.2

1	2	3	4	5	6	7	8	9
$O_2(32)$	$Ne^+(20)$	.025-.9	16,32	T1	R	144	Tab.2	279
$O_2(32)$	$Si^+(28)$	.03-.9	16,32	T1	R	144	Tab.2	279
$O_2(32)$	$Kr^+(84)$	$.033\text{-}.093^2$	16	T1	ND	77	Fig.2	1153
$O_2(32)$	$Kr^+(84)$	.035-.9	16,32	T1	R	144	Tab.2	279
$O_2(32)$	$Xe^+(131)$	$.033\text{-}.093^2$	16	T1	ND	77	Fig.2	1153
$O_2(32)$	$Xe^+(131)$	.04-.9	16,32	T1	R	144	Tab.2	279
$O_2(32)$	$H(1)$	7-30	$16,32^3$	T1	R	66	Fig.7	271
$O_2(32)$	$H_2^+(2)$	5-180	$16,16^{+2},32$	T1	A	141	Fig.9	32
n=2	**q= -1**							
$NaI^-(150)$	$Ar(40)$	?	127^-	P1	?	145	text	6
$Sb_2^-(244)$	$Ar(40)$	5-175	$122,122^-$	P1	A	146	Text	310
$Sb_2^-(244)$	$N_2(28)$	5-175	$122,122^-$	P1	A	146	Text	310
$Te_2^-(256)$	$He(4)$	0.3-0.8	128^-	P1	R	145	Fig.2	
$Te_2^-(256)$	$Ar(40)$	0.3-0.8	128^-	P1	R	145	Fig.2	
$Bi_2^-(418)$	$He(4)$	1.1	209^-	P1	?	145	text	5
n=3	**q=2**							
$CO_2^{+2}(44)$	$He(4)$	17.5-30	16	P3	R	31	Tab.4	198
$SF_2^{+2}(70)$	$He(4)$	30-70	19	P3	R	31	Tab.4	198

52

TABLE 4.2

1	2	3	4	5	6	7	8	9
n=3	q=1							
$H_3^+(3)$	He(4)	2-50	1,2	P1	ND	25	Fig.9	68
$H_3^+(3)$	He(4)	3-50	1,2,1$^-$	P1	A	91	Tab.14	169
$H_3^+(3)$	Ne(20)	2-50	1,2	P1	ND	25	Fig.9	68
$H_3^+(3)$	Ne(20)	3-50	1,2,1$^-$	P1	A	91	Tab.14	169
$H_3^+(3)$	Ne(20)	5-25	1,2	P1	A	23	Fig.11	
$H_3^+(3)$	Ar(40)	2-50	1,2	P1	ND	25	Fig.9	68
$H_3^+(3)$	Ar(40)	3-50	1,2,1$^-$	P1	A	91	Tab.14	169
$H_3^+(3)$	Ar(40)	5-25	1,2	P1	A	23	Fig.11	
$H_3^+(3)$	Xe(131)	60,200	1,2	P2	A	115	Tab.6.2	81
$H_3^+(3)$	$H_2(2)$	2-50	1,2	P1	ND	25	Fig.8	68
$H_3^+(3)$	$H_2(2)$	5,12,20	2,1,1$^-$	P1c	ND	25	Tab.3	68
$H_3^+(3)$	$H_2(2)$	5-25	1,2	P1	A	23	Fig.11	
$H_3^+(3)$	$H_2(2)$	5-50	1	P1c	A	109	Fig.4	
$H_3^+(3)$	$H_2(2)$	5-50	2	P1c	A	109	Fig.3	
$H_3^+(3)$	$H_2(2)$	5-120	1,2,1°,2°	P1c	ND	27	Fig.10	1858
$H_3^+(3)$	$H_2(2)$	60,200	1,2	P2	A	115	Tab.6.2	81
$H_3^+(3)$	$H_2(2)$	5-13	1	P3	R	112	Fig.2	177
$H_3^+(2)$	$H_2(2)$	40-220	1,2	P5	A	115	Fig.6.2	81
$H_3^+(3)$	$N_2(28)$	5-25	1,2	P1	A	23	Fig.11	
$H_3^+(3)$	$H_2O(18)$	60	1°	P2	R	147	Fig.5	
$H_3^+(3)$	$H_2O(18)$	60,200	1,2	P2	A	115	Tab.6.2	81
$D_3^+(6)$	$D_2(4)$	30-100	2,4	P1	A	148	Tab.	540
$CH_2^+(14)$	He(4)	2.5	12	P1	R	22	Tab.2	84

TABLE 4.2

1	2	3	4	5	6	7	8	9
$CH_2^+(14)$	Ne(20)	2.5	12	P1	R	22	Tab.2	84
$CH_2^+(14)$	Ne(20)	3.5	12,13	P1a	A	125	Text,Fig.	1499
$CH_2^+(14)$	Ar(40)	2.0-3.0	12-13	P1	R	149	Tab.1,2,3	10
$CH_2^+(14)$	Ar(40)	2.5	12	P1	R	22	Tab.2	84
$CH_2^+(14)$	H_2(2)	2.0-3.0	12-13	P1	R	149	Tab.1,2,3	10
$CH_2^+(14)$	H_2(2)	2.5	12	P1	R	22	Tab.2	84
$CH_2^+(14)$	D_2(4)	3-60	12,13	P1	A	150	Fig.6	621
$CH_2^+(14)$	CH_4^+(16)	2.0-3.0	12-13	P1	R	149	Tab.1,2,3	10
$CH_2^+(14)$	Air(~30)	2.0-3.0	12-13	P1	R	149	Tab.1,2,3	10
$CH_2^+(14)$	Air(~30)	2.5	12	P1	R	8	Tab.	460
$CH_2^+(14)$	Air(~30)	3-100	12,13	P1	A	150	Fig.6	621
$CH_2^+(14)$	Air(~30)	3.5	12-13	P1	A	126	Tab.3	1077
$CH_2^+(14)$	Air(~30)	3.5,90	12,13	P1	R	62	Fig.2	1859
$CH_2^+(14)$	$n\text{-}C_4H_{10}$(58)	1.3-3.0	12-13	P1	R	149	Tab.1,2,4	
$CH_2^+(14)$	$i\text{-}C_4H_{10}$(58)	1.3-3.0	12-13	P1	R	149	Tab.1,2,4	
$CH_2^+(14)$	$n\text{-}C_4H_{10}$(58)	2.5	12	P1	R	22	Tab.2	84
$CH_2^+(14)$	$n\text{-}C_8H_{18}$(114)	2.5	12	P1	R	22	Tab.2	84
$CH_2^+(14)$	$CHCl_3$(118)	2.5	12-13	P1	R	149	Tab.2	
$CH_2^+(14)$	$CHBr_3$(253)	1.3-3.0	12-13	P1	R	149	Tab.1,3,4	
$CH_2^+(14)$	$(CHCl)_2$(96)	3.0	12-13	P1	R	149	Tab.1	
$CH_2^+(14)$	$CHCl_2CH_3$(98)	2.5	12-13	P1	R	149	Tab.2	
$CH_2^+(14)$	$n\text{-}C_8H_{18}$(114)	2.5	12-13	P1	R	149	Tab.2	
$NH_2^+(16)$	Air(~30)	3-80	1,14,15	P1	A	150	Fig.8	622
$NH_2^+(16)$	Air(~30)	3.5	14-15	P1	A	126	Tab.3	1077

TABLE 4.2

1	2	3	4	5	6	7	8	9
$H_2O^+(18)$	Air(~30)	3-60	1,16,17	P1	A	150	Fig.8	622
$H_2O^+(18)$	Air(~30)	3.5	16,17	P1	R	126	Tab.1	1074
$H_2O^+(18)$	Air(~30)	3.5	17,16	P1	A	126	Tab.3	1077
$C_2H^+(25)$	Ne(20)	3.5	12,13,24	P1	A	126	Tab.3	1077
$C_2H^+(25)$	Ne(20)	3.5	12-24	P1	A	127	Tab.1	1019
$C_2H^+(25)$	Air(~30)	3.5	12,13,24	P1	A	126	Tab.3	1077
$C_3^+(36)$	Ne(20)	3.5	24	P1	A	127	Tab.1	1019
$CO_2^+(44)$	Ne(40)	1.65	12,16,28	P1	R	15	Tab.2	208
$CO_2^+(44)$	Air(~30)	2.5	12,16	P1	R	8	Tab.	460
$CO_2^+(44)$	He(4)	30-70	16	P3	R	31	Tab.4	198
$N_2O^+(44)$	$N_2O(46)$	1.0	30	P1	ND	17	Text	2270
$SO_2^+(64)$	He(4)	17.5-30	16	P3	R	31	Tab.4	198
n=3	**q=0**							
$H_2O(18)$	e	0.5	16-18	T1	ND	151	Tab.2	539
$H_2O(18)$	$Ar^{+2}(40)$	0.5	16-18	T1	ND	151	Tab.2	539
$H_2O(18)$	$Kr^{+2}(84)$	0.5	16-18	T1	ND	151	Tab.2	539
$H_2O(18)$	$Xe^{+2}(131)$	.017-.9	16-19	T1	R	152	Tab.2	539
$H_2O(18)$	$B^+(11)$	0.5	16-18	T1	ND	151	Tab.2	539
$H_2O(18)$	$C^+(12)$	0.5	16-18	T1	ND	151	Tab.2	539
$H_2O(18)$	$N^+(14)$	0.5	16-18	T1	ND	151	Tab.2	539
$H_2O(18)$	$O^+(16)$	0.5	16-18	T1	ND	151	Tab.2	539
$H_2O(18)$	$F^+(19)$	.010-.9	16-19	T1	R	152	Tab.2	539

TABLE 4.2

1	2	3	4	5	6	7	8	9
$H_2O(18)$	$F^+(19)$	0.5	16-18	T1	ND	151	Tab.2	539
$H_2O(18)$	$Ne^+(20)$	.025-.9	16-19	T1	R	152	Tab.2	539
$H_2O(18)$	$Ne^+(20)$	0.5	16-18	T1	ND	151	Tab.2	539
$H_2O(18)$	$P^+(31)$	0.5	16-18	T1	ND	151	Tab.2	539
$H_2O(18)$	$S^+(32)$	0.5	16-18	T1	ND	151	Tab.2	539
$H_2O(18)$	$Cl^+(35)$	0.5	16-18	T1	ND	151	Tab.2	539
$H_2O(18)$	$Ar^+(40)$	.012-.9	16-19	T1	R	152	Tab.2	539
$H_2O(18)$	$Ar^+(40)$	0.5	16-18	T1	ND	151	Tab.2	539
$H_2O(18)$	$Zn^+(65)$	0.5	16-18	T1	ND	151	Tab.2	539
$H_2O(18)$	$Se^+(79)$	0.5	16-18	T1	ND	151	Tab.2	539
$H_2O(18)$	$Br^+(80)$	0.5	16-18	T1	ND	151	Tab.2	539
$H_2O(18)$	$Kr^+(84)$	.011-.9	16-19	T1	R	152	Tab.2	539
$H_2O(18)$	$Kr^+(84)$	0.5	16-18	T1	ND	151	Tab.2	539
$H_2O(18)$	$H_2O^+(18)$	0.5	16-18	T1	ND	151	Tab.2	539
$H_2O(18)$	$CO_2^+(44)$	.022-.9	16-19	T1	R	152	Tab.2	539
$H_2S(34)$	e	0.5	32,33,34	T1	ND	151	Tab.3	541
$H_2S(34)$	$He^{+2}(4)$	3200	1-38	T1	R	95	Tab.	83
$H_2S(34)$	$Kr^{+2}(84)$	0.5	32,33,34	T1	ND	153	Tab.	1069
$H_2S(34)$	$Kr^{+2}(84)$	0.5	32,33,34	T1	ND	151	Tab.3	541
$H_2S(34)$	$Sb^{+2}(122)$	0.5	32,33,34	T1	ND	153	Tab.	1069
$H_2S(34)$	$B^+(11)$	0.5	32,33,34	T1	ND	153	Tab.	1069
$H_2S(34)$	$B^+(11)$	0.5	32,33,34	T1	ND	151	Tab.3	541
$H_2S(34)$	$C^+(12)$	0.5	32,33,34	T1	ND	153	Tab.	1069

TABLE 4.2

1	2	3	4	5	6	7	8	9
$H_2S(34)$	$C^+(12)$	0.5	32,33,34	T1	ND	151	Tab.3	541
$H_2S(34)$	$N^+(14)$	0.5	32,33,34	T1	A	153	Tab.	1069
$H_2S(34)$	$N^+(14)$	0.5	32,33,34	T1	ND	151	Tab.3	541
$H_2S(34)$	$O^+(16)$	0.5	32,33,34	T1b	ND	153	Tab.	1069
$H_2S(34)$	$O^+(16)$	0.5	32,33,34	T1	ND	151	Tab.3	541
$H_2S(34)$	$F^+(19)$	0.5	32,33,34	T1	ND	153	Tab.	1069
$H_2S(34)$	$F^+(19)$	0.5	32,33,34	T1	ND	151	Tab.3	541
$H_2S(34)$	$Ne^+(20)$	0.5	32,33,34	T1	ND	153	Tab.	1069
$H_2S(34)$	$Ne^+(20)$	0.5	32,33,34	T1	ND	151	Tab.3	541
$H_2S(34)$	$P^+(31)$	0.5	32,33,34	T1	ND	153	Tab.	1069
$H_2S(34)$	$P^+(31)$	0.5	32,33,34	T1	ND	151	Tab.3	541
$H_2S(34)$	$S^+(32)$	0.5	32,33,34	T1	ND	153	Tab.	1069
$H_2S(34)$	$S^+(32)$	0.5	32,33,34	T1	ND	151	Tab.3	541
$H_2S(34)$	$Cl^+(35)$	0.5	32,33,34	T1	ND	153	Tab.	1069
$H_2S(34)$	$Cl^+(35)$	0.5	32,33,34	T1	ND	151	Tab.3	541
$H_2S(34)$	$Ar^+(40)$	0.5	32,33,34	T1	A	153	Tab.	1069
$H_2S(34)$	$Ar^+(40)$	0.5	32,33,34	T1	ND	151	Tab.3	541
$H_2S(34)$	$Cu^+(64)$	0.5	32,33,34	T1	ND	153	Tab.	1069
$H_2S(34)$	$Cu^+(64)$	0.5	32,33,34	T1	ND	151	Tab.3	541
$H_2S(34)$	$Zn^+(65)$	0.5	32,33,34	T1	ND	153	Tab.	1069
$H_2S(34)$	$Zn^+(65)$	0.5	32,33,34	T1	ND	151	Tab.3	541
$H_2S(34)$	$Se^+(79)$	0.5	32,33,34	T1	ND	153	Tab.	1069
$H_2S(34)$	$Se^+(79)$	0.5	32,33,34	T1	ND	151	Tab.3	541

TABLE 4.2

1	2	3	4	5	6	7	8	9
$H_2S(34)$	$Br^+(80)$	0.5	32,33,34	Tl	ND	153	Tab.	1069
$H_2S(34)$	$Br^+(80)$	0.5	32,33,34	Tl	ND	151	Tab.3	541
$H_2S(34)$	$Kr^+(84)$	0.5	32,33,34	Tl	ND	151	Tab.3	541
$H_2S(34)$	$Kr^+(84)$	0.5	32,33,34	Tl	A	153	Tab.	1069
$H_2S(34)$	$H_2S^+(34)$	0.5	32,33,34	Tl	ND	151	Tab.3	541
$H_2S(34)$	$H_2S^+(34)$	0.5	32,33,34	Tl	ND	153	Tab.	1069
$CO_2(44)$	e	0.5	12,16,28,44	Tl	ND	151	Tab.1	538
$CO_2(44)$	$He^{+2}(4)$	3200	12,16,28,44	Tl	R	95	Fig.8	45
$CO_2(44)$	$He^{+2}(4)$	3200	12,16,22,28 29,44.46	Tl	R	95	Tab.	77
$CO_2(44)$	$Ar^{+2}(40)$	0.5	12,16,28,44	Tl	ND	151	Tab.1	538
$CO_2(44)$	$Kr^{+2}(84)$	0.5	12,16,28,44	Tl	ND	151	Tab.1	538
$CO_2(44)$	$He^+(4)$	.018-.9	12,16,28,44	Tl	R	154	Tab.3	533
$CO_2(44)$	$He^+(4)$	5-45	$12,16,28,44, 12^{+2},16^{+2}, 44^{+2}$	Tl	ND	134	Fig.5	539
$CO_2(44)$	$B^+(11)$	0.5	12,16,28,44	Tl	ND	151	Tab.1	538
$CO_2(44)$	$C^+(12)$	0.5	12,16,28,44	Tl	ND	151	Tab.1	538
$CO_2(44)$	$N^+(14)$	0.5	12,16,28,44	Tl	ND	151	Tab.1	538
$CO_2(44)$	$O^+(16)$	0.5	12,16,28,44	Tl	ND	151	Tab.1	538
$CO_2(44)$	$F^+(19)$	.025-.9	12,16,28,44	Tl	R	154	Tab.3	533
$CO_2(44)$	$F^+(19)$	0.5	12,16,28,44	Tl	ND	151	Tab.1	538
$CO_2(44)$	$Ne^+(20)$	.009-.9	12,16,28,44	Tl	R	154	Tab.3	533
$CO_2(44)$	$Ne^+(20)$	0.5	12,16,28,44	Tl	ND	151	Tab.1	538

TABLE 4.2

1	2	3	4	5	6	7	8	9
$CO_2(44)$	$S^+(32)$	0.5	12,16,28,44	T1	ND	151	Tab.1	538
$CO_2(44)$	$Cl^+(35)$	0.5	12,16,28,44	T1	ND	151	Tab.1	538
$CO_2(44)$	$Ar^+(40)$	.013-093	12,16	T1	ND	77	Fig.7	1153
$CO_2(44)$	$Ar^+(40)$	.014-093	28	T1	ND	77	Fig.5	1153
$CO_2(44)$	$Ar^+(40)$	0.5	12,16,28,44	T1	ND	151	Tab.1	538
$CO_2(44)$	$Br^+(80)$	0.5	12,16,28,44	T1	ND	151	Tab.1	538
$CO_2(44)$	$Kr^+(84)$	.013-093	12,16	T1	ND	77	Fig.7	1153
$CO_2(44)$	$Kr^+(84)$	.014-093	28	T1	ND	77	Fig.5	1153
$CO_2(44)$	$Kr^+(84)$	0.5	12,16,28,44	T1	ND	151	Tab.1	538
$CO_2(44)$	$Xe^+(131)$	.013-093	12	T1	ND	77	Fig.7	1153
$CO_2(44)$	$Xe^+(131)$	.014-093	28	T1	ND	77	Fig.6	1153
$CO_2(44)$	$CO^+(28)$	.015-.9	12,16,28,44	T1	R	154	Tab.3	533
$CO_2(44)$	$CO_2^+(44)$	0.5	12,16,28,44	T1	ND	151	Tab.1	538
$N_2O(44)$	e	0.5	14,16,28, 30,44	T1	ND	151	Tab.6	545
$N_2O(44)$	$He^{+2}(4)$	3200	14-46	T1	R	95	Tab.	79
$N_2O(44)$	$Ar^{+2}(40)$	0.5	14,16,28, 30,44	T1	ND	151	Tab.5	543
$N_2O(44)$	$Kr^{+2}(84)$	0.5	14,16,28, 30,44	T1	ND	151	Tab.5	543
$N_2O(44)$	$H^+(1)$	8,30	44,30,28, 16,14,16$^-$	T1	R	155	Fig.1, Text	671,672
$N_2O(44)$	$B^+(11)$	0.5	14,16,28 30,44	T1	ND	151	Tab.6	545

TABLE 4.2

1	2	3	4	5	6	7	8	9
$N_2O(44)$	$C^+(12)$	0.5	14,16,28,30,44	T1	ND	151	Tab.6	545
$N_2O(44)$	$N^+(14)$	0.5	14,16,28,30,44	T1	ND	151	Tab.5	543
$N_2O(44)$	$O^+(16)$	0.5	14,16,28,30,44	T1	ND	151	Tab.5	543
$N_2O(44)$	$F^+(19)$	0.5	14,16,28,30,44	T1	ND	151	Tab.5	543
$N_2O(44)$	$Ne^+(20)$	0.5	14,16,28,30,44	T1	ND	151	Tab.5	543
$N_2O(44)$	$P^+(31)$	0.5	14,16,28,30,44	T1	ND	151	Tab.6	545
$N_2O(44)$	$S^+(32)$	0.5	14,16,28,30,44	T1	ND	151	Tab.6	545
$N_2O(44)$	$Cl^+(35)$	0.5	14,16,28,30,44	T1	ND	151	Tab.5	543
$N_2O(44)$	$Ar^+(40)$	.002-.012	30	T1	A	156	Fig.7	1795
$N_2O(44)$	$Ar^+(40)$	.002-.033	30	T1	A	156	Fig.3	1794
$N_2O(44)$	$Ar^+(40)$	0.5	14,16,28,30,44	T1	ND	151	Tab.5	543
$N_2O(44)$	$Se^+(79)$	0.5	14,16,28,30,44	T1	ND	151	Tab.6	545
$N_2O(44)$	$Br^+(80)$	0.5	14,16,28,30,44	T1	ND	151	Tab.6	545

TABLE 4.2

1	2	3	4	5	6	7	8	9
$N_2O(44)$	$Kr^+(84)$	.002-.010	16	T1	A	138	Fig.7	2180
$N_2O(44)$	$Kr^+(84)$	.002-.025	16	T1	A	138	Fig.3	2176
$N_2O(44)$	$Kr^+(84)$	.007-.042	30	T1	A	156	Fig.2	1793
$N_2O(44)$	$Kr^+(84)$	.077-.06	30	T1	A	138	Fig.4	2176
$N_2O(44)$	$Kr^+(84)$	0.5	14,16,28, 30,44	T1	ND	151	Tab.5	543
$N_2O(44)$	$H(1)$	8,30	44,30,28, $16,14,16^-$	T1	R	155	Fig.1, Text	671,672
$N_2O(44)$	$N_2O^+(44)$	0.5	14,16,28, 30,44	T1	ND	151	Tab.6	545
$SO_2(64)$	$He^{+2}(4)$	3200	16-66	T1	R	95	Tab.	80
$CS_2(76)$	$He^{+2}(4)$	3200	12-80	T1	R	95	Tab.	78
n=3	q=-1							
$NaI_2^-(277)$	$Ar(40)$	?	127^-150^-	P1	?	145	text	6
$Sb_3^-(366)$	$Ar(40)$	0.78	$122^-,244^-$	P1	R	145	text	5
n=4	q=2							
$SF_3^{+2}(89)$	$He(4)$	30-70	19	P3	R	31	Tab.4	198
n=4	q=1							
$CH_3^+(15)$	$He(4)$	2.5	12,13	P1	R	22	Tab.2	84
$CH_3^+(15)$	$Ne(20)$	2.5	12,13	P1	R	22	Tab.2	84

61

TABLE 4.2

1	2	3	4	5	6	7	8	9
CH_3^+(15)	Ne(20)	3.5	12-14	P1	A	127	Tab.1	1019
CH_3^+(15)	Ar(40)	1.3-3.0	12-14	P1	R	149	Tab.1,2,3	10
CH_3^+(15)	Ar(40)	2.5	12,13	P1	R	22	Tab.2	84
CH_3^+(15)	Air(~30)	3.5,90	12-14	P1	R	62	Fig.2	1859
CH_3^+(15)	H_2(2)	1.3-3.0	12-14	P1	R	149	Tab.1,2,3	10
CH_3^+(15)	H_2(2)	2.5	12,13	P1	R	22	Tab.2	84
CH_3^+(15)	D_2(4)	3-60	12-14	P1	A	150	Fig.5	621
CH_3^+(15)	Air(~30)	1.3-3.0	12-14	P1	R	149	Tab.1,2,3	10
CH_3^+(15)	Air(~30)	3-90	1,12-14	P1	A	150	Fig.5	621
CH_3^+(15)	Air(~30)	3.5	12-14	P1	A	126	Tab.3	1077
CH_3^+(15)	CH_4(16)	1.3-3.0	12-14	P1	R	149	Tab.1,2,3	10
CH_3^+(15)	$CHCl_3$(118)	2.5	12-14	P1	R	149	Tab.2	
CH_3^+(15)	$CHBr_3$(253)	1.3-3.0	12-14	P1	R	149	Tab.1,3,4	
CH_3^+(15)	$(CHCl)_2$(96)	3.0	12-14	P1	R	149	Tab.1	
CH_3^+(15)	$CHCl_2CH_3$(98)	2.5	12-14	P1	R	149	Tab.2	
CH_3^+(15)	$n\text{-}C_4H_{10}$(58)	2.5	12,13	P1	R	22	Tab.2	84
CH_3^+(15)	$n\text{-}C_4H_{10}$(58)	1.3-3.0	12-14	P1	R	149	Tab.1,2,4	
CH_3^+(15)	$i\text{-}C_4H_{10}$(58)	1.3-3.0	12-14	P1	R	149	Tab.1,2,4	
CH_3^+(15)	$n\text{-}C_8H_{18}$(114)	2.5	12,13	P1	R	22	Tab.2	84
CH_3^+(15)	$n\text{-}C_8H_{18}$(114)	2.5	12-14	P1	R	149	Tab.2	
NH_3^+(17)	Air(~30)	3.5	14-16	P1	R	126	Tab.1	1074
NH_3^+(17)	Air(~30)	3.5	14-16	P1	A	126	Tab.3	1077
$C_2H_2^+$(26)	He(4)	3	24	P3a	R	157	Fig.1	3468
$C_2H_2^+$(26)	He(4)	4	1	P3	R	157	Fig.11	3475

TABLE 4.2

1	2	3	4	5	6	7	8	9
$C_2H_2^+$(26)	Ne(20)	3.5	12-14,24,25	P1	A	126	Tab.3	1077
$C_2H_2^+$(26)	Ne(20)	3.5	12-25	P1	A	127	Tab.1	1019
$C_2H_2^+$(26)	Air(~30)	3.5	12-14,24,25	P1	A	126	Tab.3	1077
C_3H^+(37)	Ne(20)	3.5	12-36	P1	A	127	Tab.1	1019
n=4	**q=0**							
NH_3(17)	e	.06	14-17	T1	R	70	Tab.5	27
NH_3(17)	e	0.5	14-17	T1	ND	151	Tab.4	541
NH_3(17)	He^{+2}(4)	3200	1-18	T1	R	95	Tab.	82
NH_3(17)	He^{+2}(4)	3200	15-18	T1	R	95	Fig.13	48
NH_3(17)	Ar^{+2}(40)	0.5	14-17	T1	ND	151	Tab.4	541
NH_3(17)	Kr^{+2}(84)	0.5	14-17	T1	ND	151	Tab.4	541
NH_3(17)	Xe^{+2}(131)	.012-.9	14-18	T1	R	158	Tab.1	
NH_3(17)	He^+(4)	.003-.2	16,15	T1	R	159	Fig.3	
NH_3(17)	He^+(4)	.011-.9	14-18	T1	R	158	Tab.1	
NH_3(17)	B^+(11)	0.5	14-17	T1	ND	151	Tab.4	541
NH_3(17)	C^+(12)	.008-.9	14-18	T1	R	158	Tab.1	
NH_3(17)	C^+(12)	0.5	14-17	T1	ND	151	Tab.4	541
NH_3(17)	N^+(14)	0.5	14-17	T1	ND	151	Tab.4	541
NH_3(17)	O^+(16)	.012	14-18	T1	R	158	Tab.1	
NH_3(17)	O^+(16)	0.5	14-17	T1	ND	151	Tab.4	541
NH_3(17)	F^+(19)	.009-.9	14-18	T1	R	158	Tab.1	
NH_3(17)	F^+(19)	0.5	14-17	T1	ND	151	Tab.4	541

TABLE 4.2

1	2	3	4	5	6	7	8	9
$NH_3(17)$	$Ne^+(20)$	.003-.2	16,15	Tl	R	159	Fig.3	
$NH_3(17)$	$Ne^+(20)$	.009-.9	14-18	Tl	R	158	Tab.1	
$NH_3(17)$	$Ne^+(20)$	0.5	14-17	Tl	ND	151	Tab.4	541
$NH_3(17)$	$S^+(32)$	0.5	14-17	Tl	ND	151	Tab.4	541
$NH_3(17)$	$Cl^+(35)$	0.5	14-17	Tl	ND	151	Tab.4	541
$NH_3(17)$	$Ar^+(40)$	.003-.2	17,16	Tl	R	159	Fig.3	
$NH_3(17)$	$Ar^+(40)$	.004-.9	14-18	Tl	R	158	Tab.1	
$NH_3(17)$	$Ar^+(40)$	.3-.5	14-17	Tl	R	70	Tab.5	27
$NH_3(17)$	$Ar^+(40)$	0.5	14-17	Tl	ND	151	Tab.4	541
$NH_3(17)$	$Zn^+(65)$	0.5	14-17	Tl	ND	151	Tab.4	541
$NH_3(17)$	$Br^+(80)$	0.5	14-17	Tl	ND	151	Tab.4	541
$NH_3(17)$	$Kr^+(84)$	.003-.2	17,16	Tl	R	159	Fig.3	
$NH_3(17)$	$Kr^+(84)$	.012-.9	14-18	Tl	R	158	Tab.1	
$NH_3(17)$	$Kr^+(84)$	0.5	14-17	Tl	ND	151	Tab.4	541
$NH_3(17)$	$Xe^+(131)$	.003-.2	17	Tl	R	159	Fig.3	
$NH_3(17)$	$Xe^+(131)$	.012-.9	14-18	Tl	R	158	Tab.1	
$NH_3(17)$	$Xe^+(131)$	.3-.5	14-17	Tl	R	70	Tab.5	27
$NH_3(17)$	$H_2^+(2)$	.011-.9	14-18	Tl	R	158	Tab.1	
$NH_3(17)$	$NH^+(15)$	.010-.9	14-18	Tl	R	158	Tab.1	
$NH_3(17)$	$CO^+(28)$	.008-.9	14-18	Tl	R	158	Tab.1	
$NH_3(17)$	$N_2^+(28)$	.011-.9	14-18	Tl	R	158	Tab.1	
$NH_3(17)$	$N_2^+(28)$	.3-.5	14-17	Tl	R	70	Tab.5	27
$NH_3(17)$	$NH_2^+(16)$	.008-.9	14-18	Tl	R	158	Tab.1	
$NH_3(17)$	$H_2S^+(34)$	.009-.9	14-18	Tl	R	158	Tab.1	

TABLE 4.2

1	2	3	4	5	6	7	8	9
$NH_3(17)$	$N_2O^+(44)$	.009-.9	14-18	Tl	R	158	Tab.1	
$NH_3(17)$	$CO_2^+(44)$	.012-.9	14-18	Tl	R	158	Tab.1	
$NH_3(17)$	$COS^+(60)$	.014-.9	14-18	Tl	R	158	Tab.1	
$NH_3(17)$	$NH_3^+(17)$	0.5	14-17	Tl	ND	151	Tab.4	541
$NH_3(17)$	$NH_3^+(17)$	.3-.5	14-17	Ti	R	70	Tab.5	27
$C_2H_2(26)$	e	.055	12-27	Tl	R	160	Tab.1	53
$C_2H_2(26)$	e	0.07	12,13	Tl	R	94	Tab.13	64
$C_2H_2(26)$	e	0.07	1,12,13,24-26	Tl	R	94,161	Tab.18	27
$C_2H_2(26)$	e	0.70	1,12-14,24-27, 49-51	Tl	R	95	Tab.6	27
$C_2H_2(26)$	e	.25-2.75	1,12-14,24-27, 49-51	Tl	R	95	Tab.6	27
$C_2H_2(26)$	e	.27,1.0	1,6,8,12,13, 24-26	Tl	R	162	Tab.1	901
$C_2H_2(26)$	e	.50	1,12-14,24-27, 49-51	Tl	R	95	Tab.6	27
$C_2H_2(26)$	e	1.0	1,12-15,24-27, 37-39,48-52, 62-63,74-76	Tl	R	95	Tab.10	55
$C_2H_2(26)$	e	1.22	12,13,24-27	Tl	R	163	Tab.2	2784
$C_2H_2(26)$	e	2.0	1,12-14,24-27, 49-51	Tl	R	95	Tab.6	27
$C_2H_2(26)$	$He^{+2}(4)$	.9	12,13,24-27, 37,38,49-51	Tl	R,A	164	Tab.1	419

TABLE 4.2

1	2	3	4	5	6	7	8	9
C₂H₂(26)	H⁺(1)	50,100	12-27	T1	R	160	Tab.1	
C₂H₂(26)	He⁺²(4)	3200	1-52	T1	R	95	Tab.	95
C₂H₂(26)	He⁺²(4)	3200	12,13,24-26	T1	R	95	Tab.7	38
C₂H₂(26)	He⁺²(4)	3200	12,13,24-27, 37-39,48-52, 62-63,74-77	T1	R	95	Tab.11	58
C₂H₂(26)	He⁺²(4)	5100	25-27	T1	R	165	Tab.1	847
C₂H₂(26)	Kr⁺²(84)	.003,.9	12,13,24-27, 37,38,49-51	T1	R,A	164	Tab.1	419
C₂H₂(26)	Xe⁺²(131)	.004,.9	12,13,24-27, 37,38,49-51	T1	R,A	164	Tab.1	419
C₂H₂(26)	H⁺(1)	0.5	12,13,24-26	T1	R	94	Tab.13	53
C₂H₂(26)	H⁺(1)	0.5-1200	1,12,13,24-26	T1	R	94,161	Tab.18	64
C₂H₂(26)	He⁺²(4)	2000	1,6,8,12,13, 24-26	T1	R	162	Tab.1	901
C₂H₂(26)	H⁺(1)	500,2000	1,6,8,12,13, 24-26	T1	R	162	Tab.1	901
C₂H₂(26)	H⁺(1)	2250	12-26	T1	R	166	Tab.1	2308
C₂H₂(26)	H⁺(1)	2250	12,13,24-27	T1	R	163	Tab.2	2784
C₂H₂(26)	D⁺(2)	0.5	12,13,24-26	T1	R	94	Tab.13	53
C₂H₂(26)	D⁺(2)	0.5-200	1,12,13,24-26	T1	R	94,161	Tab.18	64
C₂H₂(26)	He⁺(4)	.002,.02,.9	12,13,24-27, 37,38,49-51	T1	R,A	164	Tab.1	419

TABLE 4.2

1	2	3	4	5	6	7	8	9
$C_2H_2(26)$	$He^+(4)$	0.5	12,13,24-26	Tl	R	94	Tab.13	53
$C_2H_2(26)$	$He^+(4)$	0.5-1200	1,12,13,24-26	Tl	R	94,161	Tab.18	64
$C_2H_2(26)$	$C^+(12)$	.3,.9	12,13,24-27, 37,38,49-51	Tl	R,A	164	Tab.1	419
$C_2H_2(26)$	$N^+(14)$	.003,.3,.9	12,13,24-27, 37,38,49-51	Tl	R,A	164	Tab.1	419
$C_2H_2(26)$	$N^+(14)$	0.5	13,24-26	Tl	R	94	Tab.13	53
$C_2H_2(26)$	$N^+(14)$	0.5-800	1,12,13,24-26	Tl	R	94,161	Tab.18	64
$C_2H_2(26)$	$O^+(16)$	.002,.9	12,13,24-27, 37,38,49-51	Tlb	R,A	164	Tab.1	419
$C_2H_2(26)$	$O^+(16)$	0.5	24-26	Tl	R	94	Tab.13	53
$C_2H_2(26)$	$F^+(19)$	.003,.9	12,13,24-27, 37,38,49-51	Tl	R,A	164	Tab.1	419
$C_2H_2(26)$	$Ne^+(20)$	.002,.03,.9	12,13,24-27, 37,38,49-51	Tl	R,A	164	Tab.1	419
$C_2H_2(26)$	$Ne^+(20)$	0.5	12,13,24-26	Tl	R	94	Tab.13	53
$C_2H_2(26)$	$Ne^+(20)$	0.5-600	1,12,13,24-26	Tl	R	94,161	Tab.18	64
$C_2H_2(26)$	$N^+(28)$	0.5	25,26	Tl	R	94	Tab.13	53
$C_2H_2(26)$	$Cl^+(35)$	.002,.01,	12,13,24-27, 37,38,49-51	Tl	R,A	164	Tab.1	419
$C_2H_2(26)$	$Ar^+(40)$	.002-050	12,13,24-26	Tl	A	156	Fig.4	1794
$C_2H_2(26)$	$Ar^+(40)$	.004,.9	12,13,24-27, 37,38,49-51	Tl	R,A	164	Tab.1	419

67

TABLE 4.2

1	2	3	4	5	6	7	8	9
C_2H_2(26)	Ar^+(40)	.012-.022	13	T1	A	156	Fig.8	1795
C_2H_2(26)	Ar^+(40)	0.5	12,13,24-26	T1	R	94	Tab.13	53
C_2H_2(26)	Ar^+(40)	0.5-200	1,12,13,24-26	T1	R	94,161	Tab.18	64
C_2H_2(26)	Br^+(80)	.002,.9	12,13,24-27, 37,38,49-51	T1	R,A	164	Tab.1	419
C_2H_2(26)	Kr^+(84)	.003,.9	12,13,24-27, 37,38,49-51	T1	R,A	164	Tab.1	419
C_2H_2(26)	I^+(127)	.002,19	12,13,24-27, 37,38,49-51	T1	R,A	164	Tab.1	419
C_2H_2(26)	Xe^+(131)	.002,.03	12,13,24-27, 37,38,49-51	T1	R,A	164	Tab.1	419
C_2H_2(26)	H_2^+(2)	.004,.9	12,13,24-27, 37,38,49-51	T1	R,A	164	Tab.1	419
C_2H_2(26)	H_2^+(2)	0.5	13,24-26	T1	R	94	Tab.13	53
C_2H_2(26)	H_2^+(2)	0.5-1200	1,12,13,24-26	T1	R	94,161	Tab.18	64
C_2H_2(26)	D_2^+(4)	0.5	12,13,24-26	T1	R	94	Tab.13	53
C_2H_2(26)	D_2^+(4)	0.5-200	1,12,13,24-26	T1	R	94,161	Tab.18	64
C_2H_2(26)	CH^+(13)	.003,.9	12,13,24-27, 37,38,49-51	T1	R,A	164	Tab.1	419
C_2H_2(26)	C_2^+(24)	.003,.9	12,13,24-27, 37,38,49-51	T1	R,A	164	Tab.1	419
C_2H_2(26)	CO^+(28)	.004,.03,.9	12,13,24-27, 37,38,49-51	T1	R,A	164	Tab.1	419

TABLE 4.2

1	2	3	4	5	6	7	8	9
$C_2H_2(26)$	$N_2^+(28)$	.003,.03,.9	12,13,24-27, 37,38,49-51	T1	R,A	164	Tab.1	419
$C_2H_2(26)$	$N_2^+(28)$	0.5-400	1,12,13,24-26	T1	R	94,161	Tab.18	64
$C_2H_2(26)$	$O_2^+(32)$	0.5	25,26	T1	R	94	Tab.13	53
$C_2H_2(26)$	$H_2O^+(18)$	.002,.9	12,13,24-27, 37,38,49-51	T1	R,A	164	Tab.1	419
$C_2H_2(26)$	$C_2H^+(25)$	.002,.9	12,13,24-27, 37,38,49-51	T1	R,A	164	Tab.1	419
$C_2H_2(26)$	$CHO^+(29)$	.002,.03,.1,.9	12,13,24-27, 37,38,49-51	T1	R,A	164	Tab.1	419
$C_2H_2(26)$	$CO_2^+(44)$	.003,.03,.1,.9	12,13,24-27, 37,38,49-51	T1	R,A	164	Tab.1	419
$C_2H_2(26)$	$N_2O^+(44)$	.030,.1,.9	12,13,24-27, 37,38,49-51	T1	R,A	164	Tab.1	419
$C_2H_2(26)$	$C_2H_2^+(26)$	.002,.9	12,13,24-27, 37,38,49-51	T1	R,A	164	Tab.1	419
$C_2H_2(26)$	$CH_3^+(15)$	.003,.9	12,13,24-27, 37,38,49-51	T1	R,A	164	Tab.1	419
$C_2H_2(26)$	$CH_4^+(16)$	.002,.9	12,13,24-27, 38,30,49-51	T1	R,A	164	Tab.1	419
$C_2H_2(26)$	$C_2H_5^+(29)$	.010,.9	12,13,24-27, 37,38,49-51	T1	R,A	164	Tab.1	419
$AsH_3(78)$	$He^{+2}(4)$	3200	1-79	T1	R	95	Tab.	84

TABLE 4.2

1	2	3	4	5	6	7	8	9
n=5	q=2							
$C_2H_3^{+2}$ (27)	Air(30)	9.0	24,25,26	P1	A	57	Text	352
SF_4^{+2} (108)	He(4)	30-70	19	P3	R	31	Tab.4	198
n=5	q=1							
CH_4^+(16)	He(4)	2.5	12-14	P1	R	22	Tab.2	84
CH_4^+(16)	He(4)	17.5-30	15	P3	R	31	Tab.4	198
CH_4^+(16)	Ne(20)	2.5	12-14	P1	R	22	Tab.2	84
CH_4^+(16)	Ar(40)	2.0-3.0	12-15	P1	R	149	Tab.1,2,3	10
CH_4^+(16)	Ar(40)	2.5	12-14	P1	R	22	Tab.2	84
CH_4^+(16)	H_2(2)	2.0-3.0	12-15	P1	R	149	Tab.1,2,3	10
CH_4^+(16)	H_2(2)	2.5	12-14	P1	R	22	Tab.2	84
CH_4^+(16)	D_2(4)	3.5-90	15,14,13,12	P1	A	126	Fig.2	1075
CH_4^+(16)	Air(~30)	1.3-3.0	12-15	P1	R	22	Fig.7	83
CH_4^+(16)	Air(~30)	2.0-3.0	12-15	P1	R	149	Tab.1,2,3	10
CH_4^+(16)	Air(~30)	3.5	12-15	P1	R	126	Tab.1	1074
CH_4^+(16)	Air(~30)	3.5	12-15	P1	A	126	Tab.3	1077
CH_4^+(16)	Air(~30)	3.5-90	15,14,13,12	P1	A	126	Fig.2	1075
CH_4^+(16)	Air(~30)	3.5,90	12-15	P1	R	62	Fig.2	1859
CH_4^+(16)	CH_4(16)	2.0-3.0	12-15	P1	R	149	Tab.1,2,3	10
CH_4^+(16)	$CHCl_3$(118)	2.5	12-15	P1	R	149	Tab.2	
CH_4^+(16)	$CHBr_3$(253)	1.3-3.0	12-15	P1	R	149	Tab.1,3,4	
CH_4^+(16)	$(CHCl)_2$(96)	3.0	12-15	P1	R	149	Tab.1	

TABLE 4.2

1	2	3	4	5	6	7	8	9
$CH_4^+(16)$	$CHCl_2CH_3(98)$	2.5	12-15	P1	R	149	Tab.2	
$CH_4^+(16)$	$n\text{-}C_4H_{10}(58)$	1.3-3.0	12-15	P1	R	149	Tab.1,2,4	
$CH_4^+(16)$	$n\text{-}C_4H_{10}(58)$	2.5	12-14	P1	R	22	Tab.2	84
$CH_4^+(16)$	$i\text{-}C_4H_{10}(58)$	1.3-3.0	12-15	P1	R	149	Tab.1,2,4	
$CH_4^+(16)$	$n\text{-}C_8H_{18}(114)$	2.5	12-15	P1	R	149	Tab.2	
$CH_4^+(16)$	$n\text{-}C_8H_{18}(114)$	2.5	12-14	P1	R	22	Tab.2	84
$CH_4^+(16)$	?	?	12-15	P1a,b	R	22	Tab.1	83
$CD_4^+(20)$	Air(~30)	2.8	12-18	P1	A	167	Tab.1	1615
$C_2H_3^+(27)$	Ne(20)	3.5	12-26	P1	A	127	Tab.1	1019
$C_2H_3^+(27)$	Ne(20)	3.5	12-15,24-26	P1	A	126	Tab.3	1077
$C_2H_3^+(27)$	Air(~30)	3.5	12-15,24-26	P1	A	126	Tab.3	1077
$C_2H_3^+(27)$	Air(~30)	3.5	24-26	P1a	R	62	Fig.4	1861
$C_2H_3^{+2}(27)$	Air(30)	9.0	24,25,26	P1	A	57	Text	352
$C_3H_2^+(38)$	Ne(20)	3.5	12-37	P1	A	127	Tab.1	1019
$HCOOH^+(46)$	He(4)	5.1	18,29,45	P1	R	168	Tab.1	404
$HCOOH^+(46)$	Ar(40)	5.1	18,29,45	P1	R	168	Tab.1	404
$HCOOH^+(46)$	Kr(84)	5.1	18,29,45	P1	R	168	Tab.1	404
$HCOOH^+(46)$	$H_2(2)$	5.1	17,18,28,29,45	P1	A	168	Tab.2	404
$HCOOH^+(46)$	$H_2(2)$	5.1	18,29,45	P1	R	168	Tab.1	404
$HCOOH^+(46)$	$D_2(4)$	5.1	18,29,45	P1	R	168	Tab.1	404
$HCOOH^+(46)$	$N_2(28)$	5.1	18,29,45	P1	R	168	Tab.1	404
$HCOOD^+(47)$	$H_2(2)$	5.1	18,28-30,45,46	P1	A	168	Tab.3	405
$DCOOH^+(47)$	$H_2(2)$	5.1	17,28-30,45,46	P1	A	168	Tab.3	405

71

TABLE 4.2

1	2	3	4	5	6	7	8	9
n=5	q=0							
$CH_4(16)$	e	.055	12-17	T1	R2	160	Tab.1	22
$CH_4(16)$	e	.060	12-16	T1	R	70	Tab.	43
$CH_4(16)$	e	0.07	12-16	T1	R	94	Tab.VIII	59
$CH_4(16)$	e	0.07	1,12-16	T1	R	94	Tab.XV	427
$CH_4(16)$	e	.075	12-16	T1	R	84	Tab.1	628
$CH_4(16)$	e	.075	12-16	T1	R	169	Tab.3	543
$CH_4(16)$	e	0.5	12-16	T1	ND	151	Tab.5	516
$CH_4(16)$	e	1.0	12-16	T1	R	83	Tab.1	50
$CH_4(16)$	e	1.0	1,12-16,17-30	T1	R	95	Tab.8	2784
$CH_4(16)$	e	1.22	12-17	T1	R	163	Tab.1	85
$CH_4(16)$	$He^{+2}(4)$	3200	1-29	T1	R	95	Tab.	53
$CH_4(16)$	$He^{+2}(4)$	3200	12-16,17,26-30	T1	R	95	Tab.9	543
$CH_4(16)$	$Ar^{+2}(40)$	0.5	12-16	T1	ND	151	Tab.5	543
$CH_4(16)$	$Kr^{+2}(84)$	0.5	12-16	T1	ND	151	Tab.5	530
$CH_4(16)$	$Xe^{+2}(131)$	.011-.9	12-17	T1	R	170	Tab.1	22
$CH_4(16)$	$H^+(1)$	.3-.5	12-16	T1	R	70	Tab.	43
$CH_4(16)$	$H^+(1)$	0.5	14-16	T1	R	94	Tab.VIII	59
$CH_4(16)$	$H^+(1)$	0.5-1200	1,12-16	T1	R	94	Tab.XV	1154
$CH_4(16)$	$H^+(1)$	5-45	1,2,12,13, 14,15,16	T1	ND	78	Fig.8	
$CH_4(16)$	$H^+(1)$	50,100	12-17	T1	R	160	Tab.1	
$CH_4(16)$	$H^+(1)$	2250	12-17	T1	R	163	Tab.1	2784

TABLE 4.2

1	2	3	4	5	6	7	8	9
CH$_4$(16)	D$^+$(2)	0.5	13-16	T1	R	94	Tab.VIII	43
CH$_4$(16)	D$^+$(2)	0.5-400	1,12-16	T1	R	94	Tab.XV	59
CH$_4$(16)	He$^+$(4)	.002-.2	12-15	T1	R	171	Fig.5	1689
CH$_4$(16)	He$^+$(4)	.003-.9	12-17	T1	R	170	Tab.1	530
CH$_4$(16)	He$^+$(4)	.030	12-17,26-29	T1	R	172	Tab.3	166
CH$_4$(16)	He$^+$(4)	.035-2.0	12-16	T1	R	83	Tab.2	516
CH$_4$(16)	He$^+$(4)	.3-.5	12-16	T1	R	70	Tab.	22
CH$_4$(16)	He$^+$(4)	0.5	12-16	T1	R	94	Tab.VIII	43
CH$_4$(16)	He$^+$(4)	0.5-1200	1,12-16	T1	R	94	Tab.XV	59
CH$_4$(16)	He$^+$(4)	1.0	12-16	T1	R	83	Tab.1	516
CH$_4$(16)	He$^+$(4)	1.0	12-16	T1	R	169	Tab.3	628
CH$_4$(16)	He$^+$(4)	5-45	1,2,12,13,14,15,16	T1	ND	134	Fig.6	540
CH$_4$(16)	B$^+$(11)	0.5	12-16	T1	ND	151	Tab.5	543
CH$_4$(16)	C$^+$(12)	.009-.9	12-17	T1	R	170	Tab.1	530
CH$_4$(16)	C$^+$(12)	0.5	12-16	T1	ND	151	Tab.5	543
CH$_4$(16)	N$^+$(14)	.003-.9	12-17	T1	R	170	Tab.1	530
CH$_4$(16)	N$^+$(14)	0.5	14-16	T1	R	94	Tab.VIII	43
CH$_4$(16)	N$^+$(14)	0.5	12-16	T1	ND	151	Tab.5	543
CH$_4$(16)	N$^+$(14)	0.5-800	1,12-16	T1	R	94	Tab.XV	59
CH$_4$(16)	O$^+$(16)	0.5	13-16	T1	R	94	Tab.VIII	43

TABLE 4.2

1	2	3	4	5	6	7	8	9
$CH_4(16)$	$O^+(16)$	0.5	12-16	T1b	ND	151	Tab.5	543
$CH_4(16)$	$F^+(19)$	.03-.9	12-17	T1b	R	170	Tab.1	530
$CH_4(16)$	$F^+(19)$	0.5	12-16	T1	ND	151	Tab.5	543
$CH_4(16)$	$Ne^+(20)$	.002-.2	13,14	T1	R	171	Fig.4	1689
$CH_4(16)$	$Ne^+(20)$	.014	12-17,26-29	T1	R	172	Tab.3	166
$CH_4(16)$	$Ne^+(20)$	.10-.9	12-17	T1	R	170	Tab.1	530
$CH_4(16)$	$Ne^+(20)$	0.5	12-16	T1	ND	151	Tab.5	543
$CH_4(16)$	$Ne^+(20)$	0.5	13-16	T1	R	94	Tab.VIII	43
$CH_4(16)$	$Ne^+(20)$	0.5-600	1,12-16	T1	R	94	Tab.XV	59
$CH_4(16)$	$Ne^+(20)$	1.0	12-16	T1	R	84	Tab.1	427
$CH_4(16)$	$Ne^+(20)$	1.0	12-16	T1	R	169	Tab.3	628
$CH_4(16)$	$Si^+(28)$	.04-.9	13-16	T1	R	144	Tab.4	282
$CH_4(16)$	$P^+(31)$	0.5	12-16	T1	ND	151	Tab.5	543
$CH_4(16)$	$S^+(32)$	0.5	12-16	T1	ND	151	Tab.5	543
$CH_4(16)$	$Cl^+(35)$	.003-.9	12-17	T1	R	170	Tab.1	530
$CH_4(16)$	$Cl^+(35)$	0.5	12-16	T1	ND	151	Tab.5	543
$CH_4(16)$	$Ar^+(40)$	.002-.2	14-16	T1	R	171	Fig.3	1689
$CH_4(16)$	$Ar^+(40)$	.003-.9	12-17	T1	R	170	Tab.1	530
$CH_4(16)$	$Ar^+(40)$	.012	12-17,26-29	T1	R	172	Tab.3	166
$CH_4(16)$	$Ar^+(40)$	.075-2.0	12-16	T1	R	83	Tab.3	517
$CH_4(16)$	$Ar^+(40)$	0.75-2.0	12-16	T1	R	83	Fig.6	518
$CH_4(16)$	$Ar^+(40)$	.3-.5	12-16	T1	R	70	Tab.	22
$CH_4(16)$	$Ar^+(40)$	0.5	14-16	T1	R	94	Tab.VIII	43
$CH_4(16)$	$Ar^+(40)$	0.5	12-16	T1	ND	151	Tab.5	543

TABLE 4.2

1	2	3	4	5	6	7	8	9
$CH_4(16)$	$Ar^+(40)$	0.5-200	1,12-16	TI	R	94	Tab.XV	59
$CH_4(16)$	$Ar^+(40)$	1.0	12-16	TI	R	83	Tab.1	516
$CH_4(16)$	$Ar^+(40)$	1.0	12-16	TI	R	84	Tab.1	427
$CH_4(16)$	$Ar^+(40)$	1.0	12-16	TI	R	169	Tab.3	628
$CH_4(16)$	$Zn^+(65)$	.3-.5	12-16	TI	R	70	Tab.	22
$CH_4(16)$	$Zn^+(65)$	0.5	12-26	TI	ND	151	Tab.5	543
$CH_4(16)$	$Se^+(79)$	0.5	12-16	TI	ND	151	Tab.5	543
$CH_4(16)$	$Br^+(80)$	0.5	12-16	TI	ND	151	Tab.5	543
$CH_4(16)$	$Kr^+(84)$	.002-.2	15,16	TI	R	171	Fig.2	1689
$CH_4(16)$	$Kr^+(84)$	.003-.9	12-17	TI	R	170	Tab.1	530
$CH_4(16)$	$Kr^+(84)$	.05-.7	13-16	TI	R	169	Fig.6	629
$CH_4(16)$	$Kr^+(84)$	0.5	12-16	TI	R	151	Tab.5	543
$CH_4(16)$	$Kr^+(84)$	0.5	14-16	TI	R	94	Tab.VIII	43
$CH_4(16)$	$Kr^+(84)$	1.0	12-16	TI	R	84	Tab.1	427
$CH_4(16)$	$Kr^+(84)$	1.0	12-16	TI	R	169	Tab.3	628
$CH_4(16)$	$Xe^+(131)$	.002-.2	15,16	TI	R	171	Fig.1	1689
$CH_4(16)$	$Xe^+(131)$	.003-.9	12-17	TI	R	170	Tab.1	530
$CH_4(16)$	$Xe^+(131)$	.3-.5	12-16	TI	R	70	Tab.	22
$CH_4(16)$	$Hg^+(200)$	.1-.9	12-17	TI	R	170	Tab.1	530
$CH_4(16)$	$Hg^+(200)$	.3-.5	12-16	TI	R	70	Tab.	22
$CH_4(16)$	He(4)	.25-2.0	12-16	TI	R	84	Tab.2	430
$CH_4(16)$	$H_2^+(2)$	0.5	14-16	TI	R	94	Tab.VIII	43
$CH_4(16)$	$H_2^+(2)$	0.5-1200	1,12-16	TI	R	94	Tab.XV	59
$CH_4(16)$	$H_2^+(2)$	1.0	12-16	TI	R	84	Tab.1	427

TABLE 4.2

1	2	3	4	5	6	7	8	9
CH$_4$(16)	H$_2^+$(2)	1.0	12-16	TI	R	169	Tab.3	628
CH$_4$(16)	H$_2^+$(2)	.3-.5	12-16	TI	R	70	Tab.	22
CH$_4$(16)	D$_2^+$(4)	0.5	13-16	TI	R	94	Tab.VIII	43
CH$_4$(16)	D$_2^+$(4)	0.5-400	1,12-16	TI	R	94	Tab.XV	59
CH$_4$(16)	CO$^+$(28)	.003-.9	12-17	TI	R	170	Tab.1	530
CH$_4$(16)	CO$^+$(28)	.04-.9	13-16	TI	R	144	Tab.4	282
CH$_4$(16)	CO$^+$(28)	1.0	12-16	TI	R	83	Tab.1	516
CH$_4$(16)	CO$^+$(28)	1.0	12-16	TI	R	169	Tab.3	628
CH$_4$(16)	N$_2^+$(28)	.005-.9	12-17	TI	R	170	Tab.1	530
CH$_4$(16)	N$_2^+$(28)	0.5	14-16	TI	R	94	Tab.VIII	43
CH$_4$(16)	N$_2^+$(28)	0.5-600	1,12-16	TI	R	94	Tab.XV	59
CH$_4$(16)	N$_2^+$(28)	1.0	12-16	TI	R	83	Tab.1	516
CH$_4$(16)	N$_2^+$(28)	1.0	12-16	TI	R	84	Tab.1	427
CH$_4$(16)	N$_2^+$(28)	1.0	12-16	TI	R	169	Tab.3	628
CH$_4$(16)	NO$^+$(30)	.3-.5	12-16	TI	R	70	Tab.	22
CH$_4$(16)	O$_2^+$(32)	0.5	14-16	TI	R	94	Tab.8	43
CH$_4$(16)	N$_2$(28)	1.5	12-16	TI	R	84	Tab.2	430
CH$_4$(16)	H$_3^+$(3)	.3-.5	12-16	TI	R	70	Tab.	22
CH$_4$(16)	N$_2$O$^+$(44)	.003-.9	12-17	TI	R	170	Tab.1	530
CH$_4$(16)	CO$_2^+$(44)	.004-.9	12-17	TI	R	170	Tab.1	530
CH$_4$(16)	CHO$^+$(29)	.009-.9	12-17	TI	R	170	Tab.1	530
CH$_4$(16)	CH$_3^+$(15)	.3-.5	12-16	TI	R	70	Tab.	22
CH$_4$(16)	CH$_3^+$(15)	0.5	12-16	TI	ND	151	Tab.5	543
CH$_4$(16)	CCl$_3^+$(117)	.3-.5	12-16	TI	R	70	Tab.	22

TABLE 4.2

1	2	3	4	5	6	7	8	9
CH$_4$(16)	NH$_3^+$(17)	.3-.5	12-16	Tl	R	70	Tab.	22
CH$_4$(16)	CH$_4^+$(16)	.3-.5	12-16	Tl	R	70	Tab.	22
CD$_4$(20)	H$^+$(1)	.007-.9	12-22	Tl	R	170	Tab.2	531
CD$_4$(20)	He$^+$(4)	.03-.9	12-22	Tl	R	170	Tab.2	531
CD$_4$(20)	C$^+$(12)	.01-.9	12-22	Tl	R	170	Tab.2	531
CD$_4$(20)	N$^+$(14)	.004-.9	12-22	Tl	R	170	Tab.2	531
CD$_4$(20)	F$^+$(19)	.03-.9	12-22	Tl	R	170	Tab.2	531
CD$_4$(20)	Ne$^+$(20)	.03-.9	12-22	Tl	R	170	Tab.2	531
CD$_4$(20)	Cl$^+$(35)	.004-.9	12-22	Tl	R	170	Tab.2	531
CD$_4$(20)	Ar$^+$(40)	.003-.9	12-22	Tl	R	170	Tab.2	531
CD$_4$(20)	Kr$^+$(85)	.002-.9	12-22	Tl	R	170	Tab.2	531
CD$_4$(20)	Xe$^+$(131)	.03-.9	12-22	Tl	R	170	Tab.2	531
CD$_4$(20)	Hg$^+$(200)	.03-.9	12-22	Tl	R	170	Tab.2	531
CD$_4$(20)	CO$^+$(28)	.004-.9	12-22	Tl	R	170	Tab.2	531
CD$_4$(20)	N$_2^+$(28)	.005-.9	12-22	Tl	R	170	Tab.2	531
CD$_4$(20)	CO$_2^+$(44)	.006-.9	12-22	Tl	R	170	Tab.2	531
CD$_4$(20)	N$_2$O$^+$(44)	.003-.9	12-22	Tl	R	170	Tab.2	531
CD$_4$(20)	CHO$^+$(29)	.002-.9	12-22	Tl	R	170	Tab.2	531
CD$_4$(20)	CH$_3^+$(15)	.004-.9	12-22	Tl	R	170	Tab.2	531
CD$_4$(20)	CH$_4^+$(16)	.005-.9	12-22	Tl	R	170	Tab.2	531
CH$_3$Cl(50)	Xe^{+2}(131)	.012	13-51	Tl	R	172	Tab.5	170
CH$_3$Cl(50)	F$^+$(19)	.014-.9	13-51	Tl	R	172	Tab.5	170
CH$_3$Cl(50)	Ne$^+$(20)	.011-.9	13-51	Tl	R	172	Tab.5	170

TABLE 4.2

1	2	3	4	5	6	7	8	9
$CH_3Cl(50)$	$Ar^+(40)$	.012-.9	13-51	T1	R	172	Tab.5	170
$CH_3Cl(50)$	$Xe^+(131)$	.011-.9	13-51	T1	R	172	Tab.5	170
$CH_3Cl(50)$	$CO^+(28)$	.012-.9	13-51	T1	R	172	Tab.5	170
$CCl_3F(136)$	$F^+(19)$	.017-.9	19-117	T1	R	144	Tab.3	280
$CCl_3F(136)$	$Ne^+(20)$	.028-.9	19-117	T1	R	144	Tab.3	280
$CH_3I(142)$	$Ne^+(20)$	.030	11-142	T1	R	172	Tab.6	171
$CH_3I(142)$	$Ar^+(40)$	.030	11-142	T1	R	172	Tab.6	171
$CH_3I(142)$	$CO^+(28)$	.012-.9	11-142	T1	R	172	Tab.6	171
$CCl_4(152)$	$N^{+2}(14)$	.006-4	35-117	T1	R	173	Fig.10	791
$CCl_4(152)$	$N^+(14)$	.003-.2	35-117	T1	R	173	Fig.7	790
$CCl_4(152)$	$Ar^+(40)$	.003-.2	35-117	T1	R	173	Fig.12	791
$CCl_4(152)$	$N_2^+(28)$	.003-.2	35-117	T1	R	173	Fig.8	791
n=6	q=1							
$CD_5^+(22)$	Ne(20)	2.8	12-20	P1	A	167	Tab.1	1615
$C_2H_4^+(28)$	He(4)	.04-.25	14	P1	A	98	Fig.16	102
$C_2H_4^+(28)$	Ne(20)	3.5	12-27	P1	A	127	Tab.1	1019
$C_2H_4^+(28)$	Ne(20)	3.5	12-15,24-27	P1	A	126	Tab.3	1077
$C_2H_4^+(28)$	Ne(20)	3.5	12-14,24-27	P1	R	126	Tab.1	1074
$C_2H_4^+(28)$	Xe(131)	5-40	27,26,25,24, 14,13,12	P1	R	174	Fig.1	447
$C_2H_4^+(28)$	$H_2(2)$	.025-.20	26,27	P1	A	98	Fig.19	107
$C_2H_4^+(28)$	$H_2(2)$	.03-.20	14,15	P1	A	98	Fig.17	105
$C_2H_4^+(28)$	Air(~30)	3.5	12-14,24-27	P1	R	126	Tab.1	1074

TABLE 4.2

1	2	3	4	5	6	7	8	9
$C_2H_4^+(28)$	Air(~30)	3.5	12-15,24-27	P1	A	126	Tab.3	1077
$C_3H_3^+(39)$	Ne(20)	3.5	12-38	P1	A	127	Tab.1	1019
n=6	q=0							
$C_2H_4(28)$	e	.055	12-15,24-29	T1	R	160	Tab.1	1009
$C_2H_4(28)$	e	.06	13-28	T1	R	175	Fig.3	
$C_2H_4(28)$	e	.060	12-28	T1	R	70	Tab.2	23
$C_2H_4(28)$	e	0.07	1,12-14,24-26	T1	R	94	Tab.XVII	62,63
$C_2H_4(28)$	e	0.07	12-14,24-28	T1	R	94	Tab.XII	51b
$C_2H_4(28)$	e	1.22	12-15,24-29	T1	R	163	Tab.3	2784
$C_2H_4(28)$	$He^{+2}(4)$	3200	1-41	T1	R	95	Tab.	91
$C_2H_4(28)$	$Ar^{+2}(40)$	.010-.9	12-15,24-29	T1	R	176	Tab.1	288
$C_2H_4(28)$	$Kr^{+2}(84)$	.003-.9	12-15,24-29	T1	R	176	Tab.1	288
$C_2H_4(28)$	$Xe^{+2}(131)$	.008-.9	12-15,24-29	T1	R	176	Tab.1	288
$C_2H_4(28)$	$H^+(1)$	.005-.9	12-15,24-29	T1	R	176	Tab.1	288
$C_2H_4(28)$	$H^+(1)$	0.5	14,24-28	T1	R	94	Tab.XII	51b
$C_2H_4(28)$	$H^+(1)$	0.5	14,25-28	T1	R	177	Fig.1	3005
$C_2H_4(28)$	$H^+(1)$	0.5-1200	1,12-14,24-28	T1	R	94	Tab.XVII	62,63
$C_2H_4(28)$	$H^+(1)$	50,100	12-15,24-29	T1	R	160	Tab.1	
$C_2H_4(28)$	$H^+(1)$	200-400	1,12-14,24-28	T1	R	178	Tab.2	29
$C_2H_4(28)$	$H^+(1)$	200-1200	1,12-14,24-28	T1	R	178	Tab.1	28
$C_2H_4(28)$	$H^+(1)$	2250	12-15,24-27	T1	R	163	Tab.3	2784
$C_2H_4(28)$	$D^+(2)$	0.5	14,24-28	T1	R	94	Tab.XII	51b

TABLE 4.2

1	2	3	4	5	6	7	8	9
$C_2H_4(28)$	$D^+(2)$	0.5-200	1,12-14,24-28	Tl	R	94	Tab.XVII	62,63
$C_2H_4(28)$	$D^+(2)$	200	1,12-14,24-28	Tl	R	178	Tab.2	29
$C_2H_4(28)$	$He^+(4)$	.010-.9	12-15,24-29	Tl	R	176	Tab.1	288
$C_2H_4(28)$	$He^+(4)$	0.5	12-14,24-28	Tl	R	94	Tab.XII	51b
$C_2H_4(28)$	$He^+(4)$	0.5	15,26-28,30	Tl	R	177	Fig.1	3005
$C_2H_4(28)$	$He^+(4)$	0.5-1200	1,12-14,24-28	Tl	R	94	Tab.XVII	62,63
$C_2H_4(28)$	$He^+(4)$	200,400	1,12-14,24-28	Tl	R	178	Tab.2	29
$C_2H_4(28)$	$B^+(11)$	.004-.9	12-15,24-29	Tl	R	176	Tab.1	288
$C_2H_4(28)$	$C^+(12)$	.004-.9	12-15,24-29	Tl	R	176	Tab.1	288
$C_2H_4(28)$	$N^+(14)$	0.5	26-28	Tl	R	94	Tab.XII	51b
$C_2H_4(28)$	$N^+(14)$	0.5-800	1,12-14,24-28	Tl	R	94	Tab.XVII	62,63
$C_2H_4(28)$	$N^+(14)$	200,400	1,12-14,24-28	Tl	R	178	Tab.2	29
$C_2H_4(28)$	$O^+(16)$	.004-.9	12-15,24-29	Tl	R	176	Tab.1	288
$C_2H_4(28)$	$O^+(16)$	0.5	14,25-28	Tl	R	94	Tab.XII	51b
$C_2H_4(28)$	$F^+(19)$	.005-.9	12-15,24,29	Tl	R	176	Tab.1	288
$C_2H_4(28)$	$Ne^+(20)$	.010-.9	12-15,24-29	Tl	R	176	Tab.1	288
$C_2H_4(28)$	$Ne^+(20)$	0.5	14,25-28	Tl	R	94	Tab.XII	51b
$C_2H_4(28)$	$Ne^+(20)$	0.5	15,26-28,30	Tl	R	177	Fig.1	3005
$C_2H_4(28)$	$Ne^+(20)$	0.5-600	1,12-14,24-28	Tl	R	94	Tab.XVII	62,63
$C_2H_4(28)$	$Ne^+(20)$	200,400	1,12-14,24-28	Tl	R	178	Tab.2	29
$C_2H_4(28)$	$S^+(32)$	.003-.9	12-15,24-29	Tl	R	176	Tab.1	288
$C_2H_4(28)$	$Cl^+(35)$	.003-.9	12-15,24-29	Tl	R	176	Tab.1	288
$C_2H_4(28)$	$Ar^+(40)$	.003-.9	12-15,24-29	Tl	R	176	Tab.1	288

TABLE 4.2

1	2	3	4	5	6	7	8	9
$C_2H_4(28)$	$Ar^+(40)$	0.5	14,25-28	T1	R	94	Tab.XII	51b
$C_2H_4(28)$	$Ar^+(40)$	0.5	15,26-28,30	T1	R	177	Fig.1	3005
$C_2H_4(28)$	$Ar^+(40)$	0.5-200	1,12-14,24-28	T1	R	94	Tab.XVII	62,63
$C_2H_4(28)$	$Ar^+(40)$	200	1,12-14,24-28	T1	R	178	Tab.2	29
$C_2H_4(28)$	$Br^+(80)$	.003-9	12-15,24-29	T1	R	176	Tab.1	288
$C_2H_4(28)$	$Kr^+(84)$	.004-9	12-15,24-29	T1	R	176	Tab.1	288
$C_2H_4(28)$	$Kr^+(84)$	0.5	26-28	T1	R	94	Tab.XII	51b
$C_2H_4(28)$	$Kr^+(84)$	0.5	15,26-28,30	T1	R	177	Fig.1	3005
$C_2H_4(28)$	$Xe^+(131)$	.002-090	13,14,26-28	T1	A	156	Fig.5	1794
$C_2H_4(28)$	$Xe^+(131)$	.003-9	12-15,24-29	T1	R	176	Tab.1	288
$C_2H_4(28)$	$Xe^+(131)$	.003-9	13,14,26-28	T1	R	176	Tab.4	295
$C_2H_4(28)$	$Xe^+(131)$	.025-25	14-28	T1	R	70	Fig.3	28
$C_2H_4(28)$	$Xe^+(131)$	.029-040	14	T1	A	156	Fig.9	1796
$C_2H_4(28)$	$Xe^+(131)$	.04,.07	13-28	T1	R	175	Fig.4	1009
$C_2H_4(28)$	$Xe^+(131)$	.25	13-28	T1	R	175	Fig.3	1009
$C_2H_4(28)$	$Xe^+(131)$	.3-.5	12-28	T1	R	70	Tab.2	23
$C_2H_4(28)$	$H_2^+(2)$	0.5	13,14,24-28	T1	R	94	Tab.XII	51b
$C_2H_4(28)$	$H_2^+(2)$	0.5-1200	1,12-14,24-28	T1	R	94	Tab.XVII	62,63
$C_2H_4(28)$	$H_2^+(2)$	200,400	1,12-14,24-28	T1	R	178	Tab.2	29
$C_2H_4(28)$	$D_2^+(4)$	0.5	14,25-28	T1	R	94	Tab.XII	51b
$C_2H_4(28)$	$D_2^+(4)$	0.5-200	1,12-14,24-28	T1	R	94	Tab.XVII	62,63
$C_2H_4(28)$	$D_2^+(4)$	200	1,12-14,24-28	T1	R	178	Tab.2	29
$C_2H_4(28)$	$CO^+(28)$	.004-.9	12-15,24-29	T1	R	176	Tab.1	288

TABLE 4.2

1	2	3	4	5	6	7	8	9
$C_2H_4(28)$	$N_2^+(28)$	.003-.9	12-15,24-29	T1	R	176	Tab.1	288
$C_2H_4(28)$	$N_2^+(28)$	0.5	26-28	T1	R	94	Tab.XII	51b
$C_2H_4(28)$	$N_2^+(28)$	0.5	15,26-28,30	T1	R	177	Fig.1	3005
$C_2H_4(28)$	$N_2^+(28)$	0.5-400	1,12-14,24-28	T1	R	94	Tab.XVII	62,63
$C_2H_4(28)$	$N_2^+(28)$	200,400	1,12-14,24-28	T1	R	178	Tab.2	29
$C_2H_4(28)$	$O_2^+(32)$	0.5	14,25-28	T1	R	94	Tab.XII	51b
$C_2H_4(28)$	$HS^+(33)$	.003-.9	12-15,24-29	T1	R	176	Tab.1	288
$C_2H_4(28)$	$H_2O^+(18)$	.004-.9	12-15,24-29	T1	R	176	Tab.1	288
$C_2H_4(28)$	$CHO^+(29)$	.01-.9	12-15,24-29	T1	R	176	Tab.1	288
$C_2H_4(28)$	$H_2S^+(34)$	.004-.9	12-15,24-29	T1	R	176	Tab.1	288
$C_2H_4(28)$	$N_2O^+(44)$	.003-.9	12-15,24-29	T1	R	176	Tab.1	288
$C_2H_4(28)$	$CO_2^+(56)$	.003-.9	12-15,24-29	T1	R	176	Tab.1	288
$C_2H_4(28)$	$COS^+(60)$	.003-.9	12-15,24-29	T1	R	176	Tab.1	288
$C_2H_4(28)$	$C_2H_2^+(26)$	.003-.9	12-15,24-29	T1	R	176	Tab.1	288
$C_2H_4(28)$	$CH_3^+(15)$	.004-.9	12-15,24-29	T1	R	176	Tab.1	288
$C_2H_4(28)$	$NH_3^+(17)$	.3-.5	12-28	T1	R	70	Tab.2	23
$C_2H_4(28)$	$C_2H_3^+(27)$	.006-.9	12-15,24-29	T1	R	176	Tab.1	288
$C_2H_4(28)$	$C_2H_4^+(28)$	.01-.9	12-15,24-29	T1	R	176	Tab.1	288
$CH_2CD_2(30)$	$He^+(4)$	.005-.9	12-17,24-29	T1	R	176	Tab.2,3	295
$CH_2CD_2(30)$	$Ne^+(20)$	.009-.9	12-17,24-29	T1	R	176	Tab.2,3	295
$CH_2CD_2(30)$	$Ar^+(40)$	.012-.9	12-17,24-29	T1	R	176	Tab.2,3	295
$CH_2CD_2(30)$	$Xe^+(131)$	.010-.9	12-17,24-29	T1	R	176	Tab.2,3	295
$CH_2CD_2(30)$	$N_2^+(28)$	.008-.9	12-17,24-29	T1	R	176	Tab.2,3	295
$CH_2CD_2(30)$	$H_2S^+(34)$	.010-.9	12-17,24-29	T1	R	176	Tab.2,3	295

TABLE 4.2

1	2	3	4	5	6	7	8	9
$CH_2CD_2(30)$	$CO_2^+(44)$	.008-.9	12-17,24-29	T1	R	176	Tab.2,3	295
$CH_2CD_2(30$	$COS^+(60)$	.010	12-17,24-29	T1	R	176	Tab.2,3	295
$N_2H_4(32)$	e	.014	14-32	T1	R	70	Tab.3	24
$N_2H_4(32)$	$NH_3^+(17)$	.3-.5	14-32	T1	R	70	Tab.3	24
$CH_3OH(32)$	e	.070	1,2,12-19, 27-34,47, 63,65	T1	R	95	Tab.12	60
$CH_3OH(32)$	e	1.0	1,2,12-19, 27-34,47, 63,65	T1	R	95	Tab.13	64
$CH_3OH(32)$	$He^{+2}(4)$	.016-.9	12-33	T1	R	179	Tab.1	99
$CH_3OH(32)$	$He^{+2}(4)$	3200	1-34	T1	R	95	Tab.	96
$CH_3OH(32)$	$He^{+2}(4)$	3200	14-16,28-33, 47,65	T1	R	95	Tab.14	64
$CH_3OH(32)$	$Ne^{+2}(20)$	.013-.9	12-33	T1	R	179	Tab.1	99
$CH_3OH(32)$	$Ar^{+2}(40)$	.019-.9	12-33	T1	R	179	Tab.1	99
$CH_3OH(32)$	$Kr^{+2}(84)$	.008-.9	12-33	T1	R	179	Tab.1	99
$CH_3OH(32)$	$I^{+2}(127)$	.021	12-33	T1	R	179	Tab.1	99
$CH_3OH(32)$	$Xe^{+2}(131)$	.009-.9	12-33	T1	R	179	Tab.1	99
$CH_3OH(32)$	$Hg^{+2}(200)$	.010-.9	12-33	T1	R	179	Tab.1	100
$CH_3OH(32)$	$H^+(1)$	.004-.9	12-33	T1	R	179	Tab.1	98
$CH_3OH(32)$	$He^+(4)$	.016-.9	12-33	T1	R	179	Tab.1	98
$CH_3OH(32)$	$B^+(11)$	.009	12-33	T1	R	179	Tab.1	98

TABLE 4.2

1	2	3	4	5	6	7	8	9
$CH_3OH(32)$	$C^+(12)$	.008-.9	12-33	T1b	R	179	Tab.1	98
$CH_3OH(32)$	$N^+(14)$	.010-.9	12-33	T1b	R	179	Tab.1	98
$CH_3OH(32)$	$O^+(16)$	.009-.012	12-33	T1b	R	179	Tab.1	98
$CH_3OH(32)$	$F^+(19)$	.009-.9	12-33	T1b	R	179	Tab.1	98
$CH_3OH(32)$	$Ne^+(20)$	.012	12-33	T1	R	179	Tab.1	99
$CH_3OH(32)$	$Si^+(28)$	.008-.9	12-33	T1	R	179	Tab.1	99
$CH_3OH(32)$	$P^+(31)$	.011-.9	12-33	T1	R	179	Tab.1	99
$CH_3OH(32)$	$S^+(32)$	.003-.9	12-33	T1	R	179	Tab.1	99
$CH_3OH(32)$	$Cl^+(35)$	.006-.9	12-33	T1	R	179	Tab.1	99
$CH_3OH(32)$	$Ar^+(40)$	.010	12-33	T1	R	179	Tab.1	99
$CH_3OH(32)$	$Br^+(80)$	.011	12-33	T1	R	179	Tab.1	99
$CH_3OH(32)$	$Kr^+(84)$	.008	12-33	T1	R	179	Tab.1	99
$CH_3OH(32)$	$I^+(127)$	.009-.9	12-33	T1	R	179	Tab.1	99
$CH_3OH(32)$	$Xe^+(131)$	.009	12-33	T1	R	179	Tab.1	99
$CH_3OH(32)$	$Hg^+(200)$	.008-.9	12-33	T1	R	179	Tab.1	99
$CH_3OH(32)$	$H_2^+(2)$	.002-.9	12-33	T1	R	179	Tab.1	100
$CH_3OH(32)$	$CH^+(13)$	.009	12-33	T1	R	179	Tab.1	100
$CH_3OH(32)$	$CO^+(28)$	.007-.9	12-33	T1b	R	179	Tab.1	100
$CH_3OH(32)$	$N_2^+(28)$	.004	12-33	T1	R	179	Tab.1	100
$CH_3OH(32)$	$NO^+(30)$	.008	12-33	T1	R	179	Tab.1	100
$CH_3OH(32)$	$SH^+(33)$	.003-.9	12-33	T1	R	179	Tab.1	100
$CH_3OH(32)$	$CH_2^+(14)$	.006-.9	12-33	T1	R	179	Tab.1	100
$CH_3OH(32)$	$CO_2^+(44)$	.005	12-33	T1	R	179	Tab.1	101

TABLE 4.2

1	2	3	4	5	6	7	8	9
$CH_3OH(32)$	$CHO^+(29)$	.006-.9	12-33	T1	R	179	Tab.1	101
$CH_3OH(32)$	$H_2S^+(34)$	.003-.9	12-33	T1	R	179	Tab.1	101
$CH_3OH(32)$	$N_2O^+(44)$	.007	12-33	T1	R	179	Tab.1	101
$CH_3OH(32)$	$CH_3^+(15)$	.002-.9	12-33	T1	R	179	Tab.1	100
$CH_3OH(32)$	$CH_4^+(16)$	.002-.9	12-33	T1	R	179	Tab.1	100
$CH_3OH(32)$	$CH_2OH^+(31)$	.012-.9	12-33	T1	R	179	Tab.1	101
$CH_3OH(32)$	$CH_3OH^+(32)$	.010-.9	12-33	T1	R	179	Tab.1	101
$CH_3OH(32)$	$CH_3OH_2^+(33)$	.008-.9	12-33	T1	R	179	Tab.1	101
$CH_3OH(32)$	$C_2H_4OH^+(45)$	.9	12-33	T1	R	179	Tab.1	101
$CH_3OH(32)$	$C_2H_5OH^+(46)$	.005-.9	12-33	T1	R	179	Tab.1	101
$n=7$	$q=1$							
$C_2H_5^+(29)$	$Ne(20)$	3.5	12-28	P1	A	127	Tab.1	1019
$C_3H_4^+(40)$	$Ne(20)$	3.5	12-39	P1	A	127	Tab.1	1019
$n=7$	$q=0$							
$C_2H_5(29)$	$Ne^+(20)$	?	26-28,41,42	T1	R	81	Tab.I	2541
$C_2H_5(29)$	$Ar^+(40)$	?	26,29,41	T1	R	81	Tab.I	2541
$C_2H_5(29)$	$Kr^+(84)$	?	26-29,41,43	T1	R	81	Tab.I	2541
$C_2H_5(29)$	$Xe^+(131)$	?	27,29	T1	R	81	Tab.I	2541
$C_2H_3D_2(31)$	$Ar^+(40)$	?	15,26-31,41-43	T1	R	81	Tab.3	2543
$C_2H_3D_2(31)$	$Kr^+(84)$	.027-.050	26-31,43,44	T1	R	81	Fig.1	2540
$C_2H_3D_2(31)$	$Kr^+(84)$	?	26-31,41-45	T1	R	81	Tab.3	2543

85

TABLE 4.2

1	2	3	4	5	6	7	8	9
$C_2H_3D_2(31)$	$Xe^+(131)$	?	28-31,43-45	T1	R	81	Tab.3	2543
$C_2D_3H_2(32)$	$Ar^+(40)$	?	18,27-32,44-46	T1	R	81	Tab.II	2542
$C_2D_3H_2(32)$	$Kr^+(84)$	?	27-32,44-49	T1	R	81	Tab.II	2542
$C_2D_3H_2(32)$	$Xe^+(131)$	?	28-32,47-49	T1	R	81	Tab.II	2542
$CH_3NH_2(31)$	$He^+(4)$	.004-.9	14-18,26-32	T1	R	180	Tab.1	568
$CH_3NH_2(31)$	$B^+(10)$	.008-.9	14-18,26-32	T1	R	180	Tab.1	568
$CH_3NH_2(31)$	$C^+(12)$	.005-.9	14-18,26-32	T1	R	180	Tab.1	568
$CH_3NH_2(31)$	$N^+(14)$	.005-.9	14-18,26-32	T1b	R	180	Tab.1	568
$CH_3NH_2(31)$	$Ne^+(20)$	.004-.9	14-18,26-32	T1	R	180	Tab.1	568
$CH_3NH_2(31)$	$S^+(32)$	.004-.9	14-18,26-32	T1b	R	180	Tab.1	568
$CH_3NH_2(31)$	$Ar^+(40)$	.003-.9	14-18,26-32	T1	R	180	Tab.1	568
$CH_3NH_2(31)$	$Kr^+(84)$	.005-.9	14-18,26-32	T1	R	180	Tab.1	568
$CH_3NH_2(31)$	$Xe^+(131)$	.005-.9	14-18,26-32	T1	R	180	Tab.1	568
$CH_3NH_2(31)$	$Hg^+(200)$	.011-.9	14-18,26-32	T1	R	180	Tab.1	568
$CH_3NH_2(31)$	$CO^+(28)$	.004-.9	14-18,26-32	T1	R	180	Tab.1	568
$CH_3NH_2(31)$	$CS_2^+(76)$	.005-.9	14-18,26-32	T1	R	180	Tab.1	568
$CH_3NH_2(31)$	$CH_3^+(15)$	.003-.9	14-18,26-32	T1	R	180	Tab.1	568
$CH_3NH_2(31)$	$C_2H_2^+(26)$	.004-.9	14-18,26-32	T1	R	180	Tab.1	568
$CH_3NH_2(31)$	$CH_4^+(16)$	.004-.9	14-18,26-32	T1	R	180	Tab.1	568
$CH_3NH_2(31)$	$C_2H_4^+(28)$	.005-.9	14-18,26-32	T1	R	180	Tab.1	568
$CH_3NH_2(31)$	$C_2H_6^+(30)$	.006-.9	14-18,26-32	T1	R	180	Tab.1	568
$CH_3NH_2(31)$	$C_5H_5N^+(79)$	.005-.9	14-18,26-32	T1	R	180	Tab.1	568
$CH_3NH_2(31)$	$C_6H_6^+(78)$	.005-.9	14-18,26-32	T1	R	180	Tab.1	568

TABLE 4.2

1	2	3	4	5	6	7	8	9
n=8	q=1							
$C_3H_5^+(41)$	Ne(20)	3.5	12-40	P1	A	127	Tab.1	1019
n=8	q=0							
$C_2H_6(30)$	e	.055	12-15,24-31	T1	R	160	Tab.1	
$C_2H_6(30)$	e	.060	12-30	T1	R	70	Tab.2	23
$C_2H_6(30)$	e	0.07	1,12-15,24-30	T1	R	94	Tab.XVI	60,61
$C_2H_6(30)$	e	0.07	13-15,24-30	T1	R	94	Tab.XI	49
$C_2H_6(30)$	e	1.22	12-16,24-31	T1	R	163	Tab.4	2785
$C_2H_6(30)$	$He^{+2}(4)$	3200	1-41	T1	R	95	Tab.	86
$C_2H_6(30)$	$H^+(1)$	0.5	15,25-30	T1	R	94	Tab.XI	49
$C_2H_6(30)$	$H^+(1)$	0.5	15,26-28,30	T1	R	177	Fig.2	3005
$C_2H_6(30)$	$H^+(1)$	0.5-1200	1,12-15,24-30	T1	R	94	Tab.XVI	60,61
$C_2H_6(30)$	$H^+(1)$	50,100	12-15,24-31	T1	R	160	Tab.1	
$C_2H_6(30)$	$H^+(1)$	2250	12-16,24-31	T1	R	163	Tab.4	2785
$C_2H_6(30)$	$D^+(2)$	0.5	14,15,25-30	T1	R	94	Tab.XI	49
$C_2H_6(30)$	$D^+(2)$	0.5-200	1,12-15,24-30	T1	R	94	Tab.XVI	60,61
$C_2H_6(30)$	$He^+(4)$	.03-.9	12-30	T1	R	181	Tab.3	567
$C_2H_6(30)$	$He^+(4)$	0.5	13-15,24-30	T1	R	94	Tab.XI	49
$C_2H_6(30)$	$He^+(4)$	0.5	15,26-28,30	T1	R	177	Fig.2	3005
$C_2H_6(30)$	$He^+(4)$	0.5-1200	1,12-15,24-30	T1	R	94	Tab.XVI	60,61
$C_2H_6(30)$	$C^+(12)$	.005-.9	12-30	T1	R	181	Tab.3	567
$C_2H_6(30)$	$N^+(14)$	.002-.9	12-30	T1	R	181	Tab.3	567

TABLE 4.2

1	2	3	4	5	6	7	8	9
$C_2H_6(30)$	$N^+(14)$	0.5	15,26-30	Tl	R	94	Tab.XI	49
$C_2H_6(30)$	$N^+(14)$	0.5-800	1,12-15,24-30	Tl	R	94	Tab.XVI	60,61
$C_2H_6(30)$	$O^+(16)$	0.5	14,15,25-30	Tl	R	94	Tab.XI	49
$C_2H_6(30)$	$F^+(19)$	.03-.9	12-30	Tl	R	181	Tab.3	567
$C_2H_6(30)$	$Ne^+(20)$	.03-.9	12-30	Tl	R	181	Tab.3	567
$C_2H_6(30)$	$Ne^+(20)$	0.5	13-15,26-30	Tl	R	94	Tab.XI	49
$C_2H_6(30)$	$Ne^+(20)$	0.5	15,26-28,30	Tl	R	177	Fig.2	3005
$C_2H_6(30)$	$Ne^+(20)$	0.5-600	1,12-15,24-30	Tl	R	94	Tab.XVI	60,61
$C_2H_6(30)$	$Ar^+(40)$	.005-.9	12-30	Tl	R	181	Tab.3	567
$C_2H_6(30)$	$Ar^+(40)$	0.5	14,15,26-30	Tl	R	94	Tab.XI	49
$C_2H_6(30)$	$Ar^+(40)$	0.5	15,26-28,30	Tl	R	177	Fig.2	3005
$C_2H_6(30)$	$Ar^+(40)$	0.5-200	1,12-15,24-30	Tl	R	94	Tab.XVI	60,61
$C_2H_6(30)$	$Kr^+(84)$	.003-.9	12-30	Tl	R	181	Tab.3	567
$C_2H_6(30)$	$Kr^+(84)$	0.5	15,26-30	Tl	R	94	Tab.XI	49
$C_2H_6(30)$	$Kr^+(84)$	0.5	15,26-28,30	Tl	R	177	Fig.2	3005
$C_2H_6(30)$	$Xe^+(131)$	.003-.9	12-30	Tl	R	181	Tab.3	567
$C_2H_6(30)$	$Xe^+(131)$	.3-.5	12-30	Tl	R	70	Tab.2	23
$C_2H_6(30)$	$H_2^+(2)$	0.5	14,15,25-30	Tl	R	94	Tab.XI	49
$C_2H_6(30)$	$H_2^+(2)$	0.5-1200	1,12-15,24-30	Tl	R	94	Tab.XVI	60,61
$C_2H_6(30)$	$D_2^+(4)$	0.5	14,15,25-30	Tl	R	94	Tab.XI	49
$C_2H_6(30)$	$D_2^+(4)$	0.5-100	1,12-15,24-30	Tl	R	94	Tab.XVI	60,61
$C_2H_6(30)$	$N_2^+(28)$	.005-.9	12-30	Tl	R	181	Tab.3	567
$C_2H_6(30)$	$N_2^+(28)$	0.5	14,15,26-30	Tl	R	94	Tab.XI	49
$C_2H_6(30)$	$N_2^+(28)$	0.5	15,26-28,30	Tl	R	177	Fig.2	3005

TABLE 4.2

1	2	3	4	5	6	7	8	9
$C_2H_6(30)$	$N_2^+(28)$	0.5-400	1,12-15,24-30	T1	R	94	Tab.16	60,61
$C_2H_6(30)$	$CO^+(28)$	.004-.9	12-30	T1	R	181	Tab.3	567
$C_2H_6(30)$	$O_2^+(32)$	.004-.9	12-30	T1	R	181	Tab.3	567
$C_2H_6(30)$	$O_2^+(32)$	0.5	15,26-30	T1	R	94	Tab.XI	49
$C_2H_6(30)$	$CO_2^+(44)$	.004-.9	12-30	T1	R	181	Tab.3	567
$C_2H_6(30)$	$N_2O^+(44)$	.005-.9	12-30	T1	R	181	Tab.3	567
$C_2H_6(30)$	$COS^+(60)$	.003-.9	12-30	T1	R	181	Tab.3	567
$C_2H_6(30)$	$NH_3^+(17)$	.3-.5	12-30	T1	R	70	Tab.2	23
$C_2D_6(36)$	$Ar^{+2}(40)$	.003-.9	12-36	T1	R	181	Tab.1	561
$C_2D_6(36)$	$Kr^{+2}(84)$	.01-.9	12-36	T1	R	181	Tab.1	561
$C_2D_6(36)$	$Xe^{+2}(131)$	.004-.9	12-36	T1	R	181	Tab.1	561
$C_2D_6(36)$	$He^+(4)$	.01-.9	12-36	T1	R	181	Tab.1	561
$C_2D_6(36)$	$B^+(11)$	.01-.9	12-36	T1	R	181	Tab.1	561
$C_2D_6(36)$	$C^+(12)$	.004-.9	12-36	T1	R	181	Tab.1	561
$C_2D_6(36)$	$N^+(14)$	.004-.9	12-36	T1	R	181	Tab.1	561
$C_2D_6(36)$	$O^+(16)$	.004-.9	12-36	T1b	R	181	Tab.1	561
$C_2D_6(36)$	$F^+(19)$	.004-.9	12-36	T1	R	181	Tab.1	561
$C_2D_6(36)$	$Ne^+(20)$	.01-.9	12-36	T1	R	181	Tab.1	561
$C_2D_6(36)$	$S^+(32)$	.003-.9	12-36	T1	R	181	Tab.1	561
$C_2D_6(36)$	$Ar^+(40)$	.003-.9	12-36	T1	R	181	Tab.1	561
$C_2D_6(36)$	$Br^+(80)$	.004-.9	12-36	T1	R	181	Tab.1	561
$C_2D_6(36)$	$Kr^+(84)$	.004-.9	12-36	T1	R	181	Tab.1	561
$C_2D_6(36)$	$Xe^+(131)$	.004-.9	12-36	T1	R	181	Tab.1	561
$C_2D_6(36)$	$Cl^+(35)$	.003-.9	12-36	T1	R	181	Tab.1	561

TABLE 4.2

1	2	3	4	5	6	7	8	9
$C_2D_6(36)$	$OH^+(17)$	.003-9	12-36	T1	R	181	Tab.1	561
$C_2D_6(36)$	$CO^+(28)$	.003-9	12-36	T1	R	181	Tab.1	561
$C_2D_6(36)$	$N_2^+(28)$	.003-9	12-36	T1	R	181	Tab.1	561
$C_2D_6(36)$	$O_2^+(32)$	.003-9	12-36	T1	R	181	Tab.1	561
$C_2D_6(36)$	$HS^+(33)$	.003-9	12-36	T1	R	181	Tab.1	561
$C_2D_6(36)$	$H_2O^+(18)$	.003-9	12-36	T1	R	181	Tab.1	561
$C_2D_6(36)$	$CHO^+(29)$	.003-9	12-36	T1	R	181	Tab.1	561
$C_2D_6(36)$	$CDO^+(30)$	.003-9	12-36	T1	R	181	Tab.1	561
$C_2D_6(36)$	$H_2S^+(34)$	.003-9	12-36	T1	R	181	Tab.1	561
$C_2D_6(36)$	$CO_2^+(44)$	.003-9	12-36	T1	R	181	Tab.1	561
$C_2D_6(36)$	$N_2O^+(44)$	.002-9	12-36	T1	R	181	Tab.1	561
$C_2D_6(36)$	$COS^+(60)$	.01-9	12-36	T1	R	181	Tab.1	561
$C_2D_6(36)$	$CH_3^+(15)$	.004-9	12-36	T1	R	181	Tab.1	561
$C_2D_6(36)$	$C_2H_2^+(26)$	.004-9	12-36	T1	R	181	Tab.1	561
$C_2D_6(36)$	$CH_4^+(16)$	.004-9	12-36	T1	R	181	Tab.1	561
$C_2D_6(36)$	$C_2H_3^+(27)$	.003-9	12-36	T1b	R	181	Tab.1	561
$C_2D_6(36)$	$C_2H_4^+(28)$	.003-9	12-36	T1	R	181	Tab.1	561
$C_2D_6(36)$	$C_2H_5^+(29)$	.011-9	12-36	T1	R	181	Tab.1	561
$C_2D_6(36)$	$C_2H_6^+(30)$	.003-9	12-36	T1	R	181	Tab.1	561
n=9	q=0							
$C_3H_6^+(42)$	$Ne(20)$	3.5	12-41	P1	A	127	Tab.1	1019
$C_3H_6(42)$	e	.050	14-44	T1	R	70	Tab.3	24

TABLE 4.2

1	2	3	4	5	6	7	8	9
C_3H_6(42)	Xe^+(131)	.3-.5	14-44	T1	R	70	Tab.3	24
C_2H_5OH(46)	e	.070	1,2,12-19, 24-33,40-49, 73-93	T1	R	95	Tab.15	70
C_2H_5OH(46)	e	1.0	1,2,12-19, 24-33,40-49, 73-93	T1	R	95	Tab.16	71
C_2H_5OH(46)	He^{+2}(4)	3200	1-44	T1	R	95	Tab.	97
C_2H_5OH(46)	He^{+2}(4)	3200	14,15,19, 26-33,41-47, 75,77,93	T1	R	95	Tab.17	72
C_2H_5OH(46)	Ne^{+2}(20)	.015-.03	12-47	T1	R	182	Tab.1	126
C_2H_5OH(46)	Ar^{+2}(40)	.03-.3	12-47	T1	R	182	Tab.1	127
C_2H_5OH(46)	Kr^{+2}(84)	.015-.030	12-47	T1	R	182	Tab.1	127
C_2H_5OH(46)	Xe^{+2}(131)	.015-.030	12-47	T1	R	182	Tab.1	127
C_2H_5OH(46)	Hg^{+2}(200)	.1	12-47	T1	R	182	Tab.1	127
C_2H_5OH(46)	H^+(1)	.005-.030	12-47	T1	R	182	Tab.1	126
C_2H_5OH(46)	He^+(4)	.022	12-47	T1	R	182	Tab.1	126
C_2H_5OH(46)	B^+(11)	.011-.9	12-47	T1	R	182	Tab.1	126
C_2H_5OH(46)	C^+(12)	.03-.9	12-47	T1	R	182	Tab.1	126
C_2H_5OH(46)	N^+(14)	.015-.9	12-47	T1b	R	182	Tab.1	126
C_2H_5OH(46)	O^+(16)	.015-.03	12-47	T1b	R	182	Tab.1	126
C_2H_5OH(46)	F^+(19)	.015-.03	12-47	T1b	R	182	Tab.1	126

TABLE 4.2

1	2	3	4	5	6	7	8	9
$C_2H_5OH(46)$	$Ne^+(20)$	.030	12-47	T1	R	182	Tab.1	126
$C_2H_5OH(46)$	$Si^+(28)$	.03	12-47	T1	R	182	Tab.1	126
$C_2H_5OH(46)$	$P^+(31)$	.011-.03	12-47	T1b	R	182	Tab.1	126
$C_2H_5OH(46)$	$S^+(32)$	.008-.03	12-47	T1	R	182	Tab.1	126
$C_2H_5OH(46)$	$Ar^+(40)$	.005-.9	12-47	T1	R	182	Tab.1	127
$C_2H_5OH(46)$	$Br^+(80)$	.03	12-47	T1	R	182	Tab.1	127
$C_2H_5OH(46)$	$Kr^+(84)$	.015	12-47	T1	R	182	Tab.1	127
$C_2H_5OH(46)$	$I^+(127)$	.003-.8	12-47	T1	R	182	Tab.1	127
$C_2H_5OH(46)$	$Xe^+(131)$	.015	12-47	T1	R	182	Tab.1	127
$C_2H_5OH(46)$	$Hg^+(200)$	.01-.9	12-47	T1	R	182	Tab.1	127
$C_2H_5OH(46)$	$H_2^+(2)$	.011	12-47	T1	R	182	Tab.1	127
$C_2H_5OH(46)$	$CH^+(13)$	.03	12-47	T1	R	182	Tab.1	127
$C_2H_5OH(46)$	$CO^+(28)$	.011-.015	12-47	T1b	R	182	Tab.1	127
$C_2H_5OH(46)$	$N_2^+(28)$	.015-.9	12-47	T1*	R	182	Tab.1	127
$C_2H_5OH(46)$	$NO^+(30)$	.03	12-47	T1	R	182	Tab.1	127
$C_2H_5OH(46)$	$SH^+(33)$	.005-.9	14-47	T1	R	179	Tab.2	102
$C_2H_5OH(46)$	$HS^+(33)$	.03	12-47	T1	R	182	Tab.1	127
$C_2H_5OH(46)$	$CH_2^+(14)$	.015	12-47	T1	R	182	Tab.1	127
$C_2H_5OH(46)$	$CO_2^+(44)$	.03	12-47	T1	R	182	Tab.1	127
$C_2H_5OH(46)$	$H_2S^+(34)$	.005-.9	14-47	T1	R	179	Tab.2	102
$C_2H_5OH(46)$	$H_2S^+(34)$	.03	12-47	T1	R	182	Tab.1	127
$C_2H_5OH(46)$	$N_2O^+(44)$	.022	12-47	T1	R	182	Tab.1	127
$C_2H_5OH(46)$	$CH_3^+(15)$	.002-.9	14-47	T1	R	179	Tab.2	102
$C_2H_5OH(46)$	$CH_3^+(15)$	.015	12-47	T1	R	182	Tab.1	127

TABLE 4.2

1	2	3	4	5	6	7	8	9
$C_2H_5OH(46)$	$CH_4^+(16)$	.003-.9	14-47	T1	R	179	Tab.2	102
$C_2H_5OH(46)$	$CH_4^+(16)$	.015	12-47	T1	R	182	Tab.1	127
n=10	**q=1**							
$C_3H_7^+(43)$	$He(4)$	41.5	15	P3	R	31	Fig.7	198
$C_3H_7^+(43)$	$Ne(20)$	3.5	12-42	P1	A	127	Tab.1	1019
n=10	**q=0**							
$C_3H_6O(58)$	e	.060	14-58	T1	R	175	Fig.1	1008
$C_3H_6O(58)$	e	.060	14-58	T1	R	175	Fig.2	1008
$C_3H_6O(58)$	e	.060	14-58	T1	R	70	Tab.4	26
$CH_3COCH_3(58)$	$He^{+2}(4)$	3200	1-60	T1	R	95	Tab.	100,101
$C_3H_6O(58)$	$Ar^+(40)$	.3-.5	14-58	T1	R	70	Tab.4	26
$C_3H_6O(58)$	$Zn^+(65)$	.3-.5	14-58	T1	R	70	Tab.4	26
$C_3H_6O(58)$	$Xe^+(131)$	.3-.5	14-58	T1	R	70	Tab.4	26
$C_3H_6O(58)$	$Xe^+(131)$	0.5	14-58	T1	R	175	Fig.1	1008
$C_3H_6O(58)$	$H_2^+(2)$	.3-.5	14-58	T1	R	70	Tab.4	26
$C_3H_6O(58)$	$CH_3^+(15)$	0.5	14-58	T1	R	175	Fig.2	1008
$C_3H_6O(58)$	$NO^+(30)$	.3-.5	14-58	T1	R	70	Tab.4	26
$C_3H_6O(58)$	$N_2^+(28)$	.3-.5	14-58	T1	R	70	Tab.4	26
$C_3H_6O(58)$	$CH_3^+(15)$	.3-.5	14-58	T1	R	70	Tab.4	26
$C_3H_6O(58)$	$NH_3^+(17)$	.3-.5	14-58	T1	R	70	Tab.4	26

94

TABLE 4.2

1	2	3	4	5	6	7	8	9
n=11	q=1							
$C_3H_8^+$(44)	He(4)	.05-.35	28,29	P1	A	98	Fig.10	87
$C_3H_8^+$(44)	Ne(20)	3.5	12-43	P1	A	127	Tab.1	1019
$C_3H_8^+$(44)	Ne(20)	3.5	15,26-29,39-43	P1	R	126	Tab.1	1074
$C_3H_8^+$(44)	Ar(40)	.040-.36	29	P1	A	129	Fig.1	2544
$C_3H_8^+$(44)	Ar(40)	.05-.35	28,29	P1	A	98	Fig.9	86
$C_3H_8^+$(44)	H_2(2)	.05-.35	28,29	P1	A	98	Fig.11	88
n=11	q=0							
C_3H_8(44)	e	.050	14-44	T1	R	70	Tab.3	24
C_3H_8(44)	e	1.22	12-16,24-30, 37-45	T1	R	163	Tab.5	2785
C_3H_8(44)	He^{+2}(4)	3200	1-45	T1	R	95	Tab.	87
C_3H_8(44)	Ar^{+2}(20)	.003-.9	13-45	T1	R	183	Tab.1	50,51
C_3H_8(44)	H_2^+(2)	.004-.9	13-45	T1	R	183	Tab.1	52,53
C_3H_8(44)	OH^+(17)	.007-.9	13-45	T1	R	183	Tab.1	52,53
C_3H_8(44)	CO^+(28)	.006-.9	13-45	T1	R	183	Tab.1	52,53
C_3H_8(44)	N_2^+(28)	.004-.9	13-45	T1	R	183	Tab.1	52,53
C_3H_8(44)	O_2^+(32)	.004-.9	13-45	T1	R	183	Tab.1	52,53
C_3H_8(44)	SH^+(33)	.004-.9	13-45	T1	R	183	Tab.1	54,55
C_3H_8(44)	CHO^+(29)	.011-.9	13-45	T1	R	183	Tab.1	54,55
C_3H_8(44)	CDO^+(30)	.004-.9	13-45	T1	R	183	Tab.1	52,53
C_3H_8(44)	COS^+(60)	.002-.9	13-45	T1	R	183	Tab.1	54,55

TABLE 4.2

1	2	3	4	5	6	7	8	9
$C_3H_8(44)$	$CH_2^+(14)$	.010-.9	13-45	T1	R	183	Tab.1	52,53
$C_3H_8(44)$	$H_2S^+(34)$	.005-.9	13-45	T1	R	183	Tab.1	54,55
$C_3H_8(44)$	$CO_2^+(44)$	.005-.9	13-45	T1	R	183	Tab.1	54,55
$C_3H_8(44)$	$N_2O^+(44)$	.004-.9	13-45	T1	R	183	Tab.1	54,55
$C_3H_8(44)$	$CH_3^+(15)$	.007-.9	13-45	T1	R	183	Tab.1	52,53
$C_3H_8(44)$	$NH_3^+(17)$	.3-.5	14-44	T1	R	70	Tab.3	24
$C_3H_8(44)$	$CH_4^+(16)$	.007-.9	13-45	T1	R	183	Tab.1	52,53
$C_3H_8(44)$	$CH_2OH(31)$	.011-.9	13-45	T1	R	183	Tab.1	52,53
$C_3H_8(44)$	$CD_2OH^+(33)$	.005-.9	13-45	T1	R	183	Tab.1	52,53
$C_3H_8(44)$	$C_2H_3^+(27)$	.006-.9	13-45	T1	R	183	Tab.1	52,53
$C_3H_8(44)$	$CH_3OH^+(32)$	.003-.9	13-45	T1	R	183	Tab.1	52,53
$C_3H_8(44)$	$C_2H_5^+(29)$	.006-.9	13-45	T1	R	183	Tab.1	52,53
$C_3H_8(44)$	$C_3H_7^+(43)$	.015-.9	13-45	T1	R	183	Tab.1	54,55
$C_3H_8(44)$	$C_3H_8^+(44)$	.006-.9	13-45	T1	R	183	Tab.1	54,55
n=12	**q=0**							
$C_4H_8(56)$ Buten-1	$He^{+2}(4)$	3200	1-83	T1	R	95	Tab.	92,93
$C_4H_8(56)$ i-Buten	$He^{+2}(4)$	3200	1-57	T1	R	95	Tab.	94
$n\text{-}C_3H_7OH(60)$	$He^{+2}(4)$	3200	1-92	T1	R	95	Tab.	98,99
$n\text{-}C_3H_7OH(60)$	$Ne^{+2}(20)$	.016-.9	14-61	T1	R	184	Tab.1	182
$n\text{-}C_3H_7OH(60)$	$Ar^{+2}(40)$	.012-.9	14-61	T1	R	184	Tab.1	182
$n\text{-}C_3H_7OH(60)$	$Kr^{+2}(84)$	.010-.9	14-61	T1	R	184	Tab.1	182

TABLE 4.2

1	2	3	4	5	6	7	8	9
n-C_3H_7OH(60)	I^{+2}(127)	.004-.9	14-61	T1	R	184	Tab.1	182
n-C_3H_7OH(60)	Xe^{+2}(131)	.012-.9	14-61	T1	R	184	Tab.1	182
n-C_3H_7OH(60)	Hg^{+2}(200)	.006-.9	14-61	T1	R	184	Tab.1	182
n-C_3H_7OH(60)	H^+(1)	.006-.9	14-16	T1	R	184	Tab.1	182
n-C_3H_7OH(60)	He^+(4)	.003-.9	14-61	T1	R	184	Tab.1	182
n-C_3H_7OH(60)	B^+(10)	.010-.9	14-61	T1	R	184	Tab.1	182
n-C_3H_7OH(60)	C^+(12)	.010-.9	14-61	T1	R	184	Tab.1	182
n-C_3H_7OH(60)	N^+(14)	.004-.9	14-61	T1	R	184	Tab.1	182
n-C_3H_7OH(60)	O^+(16)	.009-.9	14-61	T1b	R	184	Tab.1	182
n-C_3H_7OH(60)	F^+(19)	.010-.9	14-61	T1	R	184	Tab.1	182
n-C_3H_7OH(60)	Ne^+(20)	.080-.3	14-61	T1	R	184	Tab.1	182
n-C_3H_7OH(60)	S^+(32)	.003-.9	14-61	T1	R	184	Tab.1	182
n-C_3H_7OH(60)	Cl^+(35)	.010-.9	14-61	T1	R	184	Tab.1	182
n-C_3H_7OH(60)	Ar^+(40)	.010-.9	14-61	T1	R	184	Tab.1	182
n-C_3H_7OH(60)	Br^+(80)	.010-.9	14-61	T1	R	184	Tab.1	182
n-C_3H_7OH(60)	Kr^+(84)	.010-.9	14-61	T1	R	184	Tab.1	182
n-C_3H_7OH(60)	I^+(127)	.004-.9	14-61	T1	R	184	Tab.1	182
n-C_3H_7OH(60)	Xe^+(131)	.010-.9	14-61	T1	R	184	Tab.1	182
n-C_3H_7OH(60)	Hg^+(200)	.008-.9	14-61	T1	R	184	Tab.1	182
n-C_3H_7OH(60)	H_2^+(2)	.003-.9	14-61	T1	R	184	Tab.1	182
n-C_3H_7OH(60)	CH^+(13)	.010-.9	14-61	T1	R	184	Tab.1	182
n-C_3H_7OH(60)	CO^+(28)	.008-.9	14-61	T1	R	184	Tab.1	182
n-C_3H_7OH(60)	N_2^+(28)	.002-.9	14-61	T1	R	184	Tab.1	183
n-C_3H_7OH(60)	SH^+(33)	.003-.9	14-61	T1	R	184	Tab.1	183

TABLE 4.2

1	2	3	4	5	6	7	8	9
n-C_3H_7OH(60)	CH_2^+(14)	.010-.9	14-61	T1	R	184	Tab.1	183
n-C_3H_7OH(60)	H_2O^+(18)	.005-.9	14-61	T1	R	184	Tab.1	183
n-C_3H_7OH(60)	H_2S^+(34)	.003-.9	14-61	T1	R	184	Tab.1	183
n-C_3H_7OH(60)	N_2O^+(44)	.008-.9	14-61	T1	R	184	Tab.1	183
n-C_3H_7OH(60)	CO_2^+(44)	.010-.9	14-61	T1	R	184	Tab.1	183
n-C_3H_7OH(60)	CS_2^+(76)	.003-.9	14-61	T1	R	184	Tab.1	183
n-C_3H_7OH(60)	CH_3^+(15)	.003-.9	14-61	T1	R	184	Tab.1	183
n-C_3H_7OH(60)	CH_4^+(16)	.003-.9	14-61	T1	R	184	Tab.1	183
n-C_3H_7OH(60)	CH_2OH^+(31)	.003-.9	14-61	T1	R	184	Tab.1	183
n-C_3H_7OH(60)	$C_3H_2^+$(43)	.007-.9	14-61	T1	R	184	Tab.1	183
n-C_3H_7OH(60)	$C_3H_6OH^+$(59)	.009-.9	14-61	T1	R	184	Tab.1	183
n-C_3H_7OH(60)	$C_3H_7OH^+$(60)	.002-.9	14-61	T1	R	184	Tab.1	183
n-C_3H_7OH(60)	$C_6H_6^+$(78)	.1-.9	14-61	T1	R	184	Tab.1	183
n=14	**q=0**							
n-C_4H_{10}(58)	e	1.22	12-59	T1	R	163	Tab.6	2785
n-C_4H_{10}(58)	He^{+2}(4)	3200	1-59	T1	R	95	Tab.	88,89
n-Butan								
C_4H_{10}(58)	He^{+2}(4)	3200	1-59	T1	R	95	Tab.	90
i-Butan								
C_4H_{10}(58)	Ne^{+2}(20)	.9	14-58	T1	R	185	Tab.1	350
C_4H_{10}(58)	Ar^{+2}(40)	.009-.9	14-58	T1	R	185	Tab.1	350
C_4H_{10}(58)	Kr^{+2}(84)	.030-.9	14-58	T1	R	185	Tab.1	350
C_4H_{10}(58)	Xe^{+2}(131)	.010-.9	14-58	T1	R	185	Tab.1	351

97

TABLE 4.2

1	2	3	4	5	6	7	8	9
$C_4H_{10}(58)$	$H^+(1)$	.004-.9	14-58	T1	R	185	Tab.1	350
$C_4H_{10}(58)$	$He^+(4)$	.005-.9	14-58	T1	R	185	Tab.1	350
$C_4H_{10}(58)$	$B^+(11)$	.002-.9	14-58	T1	R	185	Tab.1	350
$C_4H_{10}(58)$	$C^+(12)$	.005-.9	14-58	T1	R	185	Tab.1	350
$C_4H_{10}(58)$	$N^+(14)$	.005-.9	14-58	T1	R	185	Tab.1	350
$C_4H_{10}(58)$	$O^+(16)$	.010-.9	14-58	T1b	R	185	Tab.1	350
$C_4H_{10}(58)$	$Ne^+(20)$	.012-.9	14-58	T1	R	185	Tab.1	350
$C_4H_{10}(58)$	$Cl^+(35)$	.015-.9	14-58	T1	R	185	Tab.1	350
$C_4H_{10}(58)$	$Ar^+(40)$	.005-.9	14-58	T1	R	185	Tab.1	350
$C_4H_{10}(58)$	$Br^+(80)$	.002-.9	14-58	T1	R	185	Tab.1	350
$C_4H_{10}(58)$	$Kr^+(84)$	.002-.9	14-58	T1	R	185	Tab.1	350
$C_4H_{10}(58)$	$I^+(127)$	.005-.9	14-58	T1	R	185	Tab.1	350
$C_4H_{10}(58)$	$Xe^+(131)$	.006-.9	14-58	T1	R	185	Tab.1	350
$C_4H_{10}(58)$	$Hg^+(200)$	.003-.9	14-58	T1	R	185	Tab.1	350
$C_4H_{10}(58)$	$H_2^+(2)$	.005-.9	14-58	T1	R	185	Tab.1	351
$C_4H_{10}(58)$	$CH^+(13)$	.005-.9	14-58	T1	R	185	Tab.1	351
$C_4H_{10}(58)$	$OH^+(17)$	.002-.9	14-58	T1	R	185	Tab.1	351
$C_4H_{10}(58)$	$CO^+(28)$	.005-.9	14-58	T1	R	185	Tab.1	351
$C_4H_{10}(58)$	$N_2^+(28)$	.009-.9	14-58	T1	R	185	Tab.1	351
$C_4H_{10}(58)$	$O_2^+(32)$	.004-.9	14-58	T1	R	185	Tab.1	351
$C_4H_{10}(58)$	$SH^+(33)$	.005-.9	14-58	T1	R	185	Tab.1	351
$C_4H_{10}(58)$	$H_3^+(3)$	.005-.9	14-58	T1	R	185	Tab.1	351
$C_4H_{10}(58)$	$CH_2^+(14)$	.012-.9	14-58	T1	R	185	Tab.1	351
$C_4H_{10}(58)$	$H_2O^+(18)$	.002-.9	14-58	T1	R	185	Tab.1	351
$C_4H_{10}(58)$	$CHO^+(29)$	.004-.9	14-58	T1	R	185	Tab.1	351

TABLE 4.2

1	2	3	4	5	6	7	8	9
$C_4H_{10}(58)$	$CDO^+(30)$	.002-.9	14-58	Tl	R	185	Tab.1	351
$C_4H_{10}(58)$	$H_2S^+(34)$	.006-.9	14-58	Tl	R	185	Tab.1	352
$C_4H_{10}(58)$	$N_2O^+(44)$	.030-.9	14-58	Tl	R	185	Tab.1	352
$C_4H_{10}(58)$	$CO_2^+(44)$	.021-.9	14-58	Tl	R	185	Tab.1	352
$C_4H_{10}(58)$	$COS^+(60)$	.008-.9	14-58	Tl	R	185	Tab.1	352
$C_4H_{10}(58)$	$CH_3^+(15)$	.009-.9	14-58	Tl	R	185	Tab.1	351
$C_4H_{10}(58)$	$C_2H_2^+(26)$	.005-.9	14-58	Tl	R	185	Tab.1	351
$C_4H_{10}(58)$	$CH_4^+(16)$	.012-.9	14-58	Tl	R	185	Tab.1	351
$C_4H_{10}(58)$	$CD_2OH^+(33)$	.002-.9	14-58	Tl	R	185	Tab.1	351
$C_4H_{10}(58)$	$C_3H_8^+(44)$	.005-.9	14-58	Tl	R	185	Tab.1	352
$C_4H_{10}(58)$	$C_4H_{10}^+(58)$	.003-.9	14-58	Tl	R	185	Tab.1	352
$n\text{-}C_4H_{10}(58)$	$H(1)$	2250	12-59	Tl	R	163	Tab.6	2785
$B_5H_9(64)$	$D^+(2)$	.002-140	46-64	Tl	R	186	Tab.2	
$B_5H_9(64)$	$He^+(4)$	.01	34-62	Tl	R	187	Tab.1	406
$B_5H_9(64)$	$He^+(4)$	.002-14	44-59	Tl	R	97	Tab.	
$B_5H_9(64)$	$Ne^+(20)$	.002-.2	44-59	Tl	R	97	Tab.	
$B_5H_9(64)$	$Ne^+(20)$	.01	34-62	Tl	R	187	Tab.1	406
$B_5H_9(64)$	$Ar^+(40)$	.006-.2	46-61	Tl	R	97	Tab.	
$B_5H_9(64)$	$Ar^+(40)$	.006-.2	46-63	Tl	R	187	Fig.1	407
$B_5H_9(64)$	$Ar^+(40)$	.01	34-62	Tl	R	187	Tab.1	406
$B_5H_9(64)$	$Kr^+(84)$	.002-.2	48-61	Tl	R	97	Tab.	
$B_5H_9(64)$	$Kr^+(84)$	.01	34-62	Tl	R	187	Tab.1	406
$B_5H_9(64)$	$Xe^+(131)$	.002-.2	48-62	Tl	R	97	Tab.	
$B_5H_9(64)$	$Xe^+(131)$	.01	34-62	Tl	R	187	Tab.1	406
$B_5H_9(64)$	$D_2^+(4)$	.002-.20	24-64	Tl	R	186	Tab.3	406

CHAPTER 5

EVALUATION CRITERIA FOR MEASUREMENTS

5.1 SIGNIFICANCE OF CRITERIA

In this section we present a list of "criteria" whose consideration is deemed essential to a well-defined dissociative collision measurement. Each criterion is important because it relates to the definition of the measured quantity or because it refers to a possible source of systematic error whose presence should be investigated and whose contribution to the overall uncertainty should be assigned a bound. Without a specific listing of the error sources that an author has considered, and a description of the manner in which the contributions from the various sources were combined in the overall error estimate, a paper is incomplete. No investigator is entirely circumspect in identifying error sources, and his mood may vary with time. It is important to report exactly what the error analysis consists of, whatever the author's disposition toward thoroughness might be. Perhaps the single most significant revelation to the present authors as a result of our study in preparation for this writing was the recognition of these important and simple facts. Our perusal of the literature, including our own papers, indicates that these principles need to be reemphasized in the interest of higher quality scientific work.

We feel that in spite of the present-day editorial pressure on authors to condense the text of papers, it is an author's responsibility to insure that his results are presented in sufficient detail in regard to definition and error analysis, and that they are intelligible and subject to fair criticism, at least to a specialist in the field. It is our hope that the following compilation of criteria serves as a basic check list for experimenters and that the list will be expanded in the future to form a basis for better planning of experiments and better reporting of results.

For the present, the list serves perhaps as a basis for subjective evaluation of the existing literature.

We leave it to the reader to apply these criteria to the individual papers on dissociation in any manner or depth appropriate to his purpose. The relative weight to be given to the various criteria in arriving at an overall evaluation of a measurement depends on the purposes for which the data are to be used.

All of the criteria should be considered if absolute cross section data are required. Those criteria relating to absolute pressure and path length determinations can be ignored if one is only concerned with the shape of the cross section versus energy behavior, or the differential cross section versus angle behavior. Nearly all of the data in the dissociation literature are correct to a plus or minus factor of two, but few error estimates can be relied on because of lack of information concerning their basis.

Even the crudest data may be of value to some users. For example, if one is concerned with whether or not a certain dissociation fragment appeared in a mass spectrum at all, he does not need to consider some of the finer points of product analysis such as the exact solid angle of the fragment detector or the exact fragment energy discrimination.

5.2 BEAM SPECIFICATION

In virtually every dissociation measurement at least one of the reactants has been formed into a beam of particles moving in essentially the same direction with a fairly well-defined energy. In order to perform a well-defined cross section measurement using such a beam, a number of critical factors related to the production, identification, collimation, and detection of the beam particles need to be examined. Sections 5.2.1 through 5.2.9 deal with these factors.

5.2.1 Atomic or Molecular Constituents

If the beam is composed of atoms or atomic ions, the value of the atomic number and the value of atomic weight must be specified. When more than one isotope is present in the source from which the beam of atoms is derived and the various isotopes are ionized and accelerated through the same potential drop, the velocities are a function of mass. If the cross sections are also strongly velocity dependent, the isotopes exhibit unequal cross sections, and a precise definition of the experimental conditions requires specification of the isotopes present.

Most apparatus utilize electrostatic ion acceleration and magnetic deflector analysis of momentum-to-charge ratio. It is usually possible to design the magnetic analyzer so that it has sufficient resolution to perform isotope selection in those cases where a significant velocity difference exists between isotopes of the same energy in the beam.

If the beam is composed of molecules or molecular ions, the value of atomic number and weight of each constituent atom must be specified, and, if more than one form exists, the isomeric form of the molecule. or molecular ion must also be specified.

In atomic beams, isotopes of the same velocity may exhibit unlike differential scattering cross sections in close collisions involving appreciable scattering.

In molecular beams, isotopic and isomeric composition variations affect the vibrational and rotational energy level system. These variations may profoundly affect collisional behavior.

Another important element of beam particle specification is the charge state of the ions or molecules present in the beam. Charge state is usually inferred from knowledge of mass-to-charge ratio coupled with knowledge of the possible mass and charge values of species that can conceivably issue from the ion source employed. The investigator or critic is well-advised to be wary of possible ambiguities in species that exist with the pairs N_2^{2+} and N^+, both of which can be produced in an ion source using electron bombardment or electrical discharges. McGowan[12] has used isotopic N_2 gas composed of mass 14 and 15 isotopes to produce separable $N_{14}N_{15}^{2+}$, N_{14}^+, and N_{15}^+ ions. Another example is that of the set He^{2+}, H_2^+, and D^+, all having the same mass-to-charge ratio. These can all issue in nearly equal number from an He-filled discharge recently contaminated with H_2 and D_2 gases. In considering the possible mass species that can emerge from an ion source, it is well not to neglect consideration of ion species that can be produced through ion molecule reactions of once or twice ionized gas molecules or atoms with neutral molecules. In dealing with polyatomic gases, electron impact may produce virtually all possible dissociation fragments of the parent gas molemolecules. For example, from electron bombardment of CH_4 one may expect as products CH_4^+, CH_3^+, CH_2^+, CH^+, C^+, H_2^+, H_3^+, and H^+.

A source of ambiguity in species with mass-to-charge ratio selection is collisional change of mass-to-charge ratio in the drift space between the electrostatic accelerator and magnetic analyzer. This can cause spurious mass peaks to appear at unexpected settings of the magnetic field. This process has been discussed in Chapter 3 under the category of Aston band studies of dissociation. Fractional mass peaks can result from charge exchange as well as dissociation, and the same considerations are applicable for predicting the spurious mass peak positions.

5.2.2 Kinetic Energy of Constituents

The kinetic energy of the beam particles is usually determined on the basis of the assumption that the charge state has a known fixed value during the passage of the ion through the accelerator. This can be assured by making the gas pressure in the accelerating column sufficiently low so that no charge transfer collisions occur during acceleration, and by making sure that the mass-to-charge ratio filter is not transmitting species which have changed their mass-to-charge ratio in the drift region between the accelerator and magnetic analyzer. In most instruments the contribution to the desired beam from both of these processes is very small. A test for both is to observe the change in beam current with gas pressure in the accelerator and analyzer. The change should be negligible when the pressure is increased by, for example, a factor of two from the normal value.

Usually ions emitted from an ion source have a distribution of from 1 to 100 eV wide. After acceleration the ions exhibit the same energy spread even though their energies have been increased by the accelerator. After passage of the accelerated ions through the magnetic analyzer, the beam components having different velocities proceed in different directions. Consequently, suitably placed collimators, apertures, or slits can narrow the transmitted energy distribution. It is advisable to specify the spread in the kinetic energies of ions passing through the collision chamber and to show that measured cross section does not vary significantly for energy changes corresponding to the extreme limits of the spread. Generally, in atom-molecule collisions there is no reason to suspect fine structure in the cross section versus energy dependence which would introduce drastic cross section changes for a few percent change in beam energy. Thus for most purposes recognized at present, spreads of a few percent easily achievable with simple design constraints should suffice.

5.2.3 Internal Energy of Constituents

Owing to the possibility, which always exists in principle, that a measured cross section is initial-state dependent, a definitive measurement calls for an accompanying assessment of possible effects of such states when their presence is possible. The investigator must consider the ion source mechanics and the modification of excited state populations through spontaneous decay, or collisions during the flight of the projectiles from the ion source to the collision region of the apparatus.

Ideally one would deal always with particles in a single well-defined state or a precisely known population of states. These ideals have rarely been achieved, except in the following cases:

1. The projectile has no internal structure (for example H^+, D^+, and He^{2+} ions or other stripped nuclei).

2. The projectile has only one state (H^- ions are an example).

3. The projectile is produced in a collision process in which the available energy is sufficient to produce only one state, and this one state is not depopulated by secondary collisions in the ion source or elsewhere along the route of the ion from the ion source to the collision region.

4. The projectiles are produced in a collision process for which the resultant excited state population, the time of flight of the projectiles through every part of the apparatus, and the decay rates of all of the initial and intermediate states which become occupied in cascade decay are known. When all of these factors are subject to quantitative evaluation, one may deduce the population arriving at the collision region.

Sometimes method 4 can be utilized, together with suitably chosen reactant flight times and quenching fields, to achieve a nearly pure ground state population or a population having set bounds on the fractional content of all excited states.

It is possible in principle to utilize a beam comprising a known mixture of states whose relative concentrations are under the control of the experimenter in order to obtain cross sections for specific excited states or limits thereon. However, the control of excited state populations in collision experiments is in its infancy at the present time. A large part of the present collision literature has not dealt adequately with the question of precollision excited states of projectiles.

It is possible to create excited states in the beam ions when either the electron impact energy on the neutral gas exceeds the threshold for the excitation of a state or when the ions, subsequent to their production, achieve kinetic energies high enough to excite themselves in collisions with gas molecules in the ion source or the drift tube upstream from the collision chamber. In the latter case, the total kinetic energy in the center-of-mass system of the ion and neutral collision partners must exceed the threshold of the lowest excited state. Vibrational and rotational excitations of molecular ions are likely to be important for collisions in the energy range from thermal to 1 keV. Electronic excitation cross sections of ions in ion-molecule collisions are generally quite small at energies below 1 keV, except in resonant charge exchange processes leading to an excited product. Exothermic ion-molecule reactions involving atom interchange can also create excited reaction products.

Generally, excited states decay exponentially in time with a rate f characteristic of the wave function of the excited species and that of the lower state to which the decay is possible. Often a state may decay to any one of several lower states, in which case the net rate $f = \sum_i f_i$, where f_i is the decay rate for all states to which a transition can occur. Many electronic states have decay rates of the order of 10^8 sec^{-1} and times of flight of ions from ion source to collision chamber of the order of 10^{-8} sec, so that a substantial fraction of the initially excited ions may decay in flight to lower states. In order to calculate the population of states in a beam at the entrance to the collision chamber, it is necessary to know the population emergent from the ion source, the lifetime of each state initially excited, and the lifetime of each state above the ground state which is populated by transitions in flight from higher states.

Fairly weak electric and magnetic fields interposed along the beam path by accident or design can sometimes drastically alter the normal decay rate of a state. For example, the metastable $2S^{1/2}$ state of the H atom is perturbed by a field of the order of 500 v cm^{-1} to such an extent that the metastable decays at a rate nearly equal to that of the $2P^{1/2}$ state. The higher angular momentum states of levels having principal quantum numbers 3, 4, and 5 can be mixed with lower angular momentum states in very weak fields to change their decay rates. Although the lower excited states of a system are generally preferentially populated, the higher states should not be ignored, since they have longer lifetimes and may possess disproportionately large cross sections for certain processes such as stripping or inducing excitation or ionization in a target molecule. Many of the above factors have been considered in more detail by McClure[188] in an experimental study of ionization and electron transfer in collisions of two H atoms, one of which was electronically excited. In this work special steps were taken to remove excited states because of indirect evidence that the ionization cross section might vary as the square of the principal quantum number, and strongly influence the results.

Molecule-ions such as H_2^+ and CO^+ are usually produced in a number of different vibrationally and rotationally excited states in electron bombardment ion sources. This circumstance is a by-product of the practical consideration that the electron energy required to ionize the parent gases H_2 and CO, at peak efficiency, and therefore to do the experiment in minimum time, is of the order of 100 eV. This energy exceeds the threshold energy for many vibrational states of the molecular ion species H_2^+ and CO^+ produced.

Since the vibrational and rotational levels of molecules are closely spaced, relatively distant collisions of the ions within the ion source can cause transitions. Thus the original population produced by electron impact may be modified substantially by collisions of the ions enroute to the exit aperture of the ion source. This effect always occurs at sufficiently high ion source pressures. Gross effects of ion source pressure on the dissociation cross section have been observed by McClure with a Penning ion source[27] and Williams with a radio frequency discharge ion source[25] in studies of the dissociation of H_2^+ projectiles incident on H_2 gas. Unfortunately, neither of these experiments was performed so that it could be established that the ionizing electron spectrum was unmodified by the pressure changes employed. It is known from other evidence (Tunitskii et al.,[7] McGowan and Kerwin,[10] Kupriyanov,[8,64,125,126] and Caudano and Delfosse[107]) that the ionizing electron spectrum can strongly influence the dissociation cross section. Tunitskii et al.,[7] and McGowan and Kerwin[10] discuss a theoretical interpretation of their results predicated on a varying vibrational population of the ions. Further information on ion source effects and excited states is referenced in Chapters 6 and 7.

A secondary collision process that may modify the vibrational population of molecular ions in an ion source is the resonant charge exchange of singly ionized parent molecules with neutral molecules. Generally, this process has a high cross section at low ion energies. Its result is the production of a set of ions that may have a modified vibrational population and a lower average energy of emergence from the ion source.

Excited vibrational and rotational states of the lowest electronic states of homopolar molecules have no dipole moment and therefore have very long lifetimes—so long, in fact, that the population is unmodified during flight from the ion source to the collision chamber in apparatus of typical dimensions. The use of heteropolar species is not expected to alter this condition.

Flight paths as long as 110 cm have been used between a neutral projectile source and collision chamber in order to permit decay in flight of excited species.[188] In the Cermak-Herman-type apparatus,[77,80] a distance of only a few centimeters between the production point and interaction region is encountered. In that type of apparatus excited states of normal (10^{-8} sec) lifetime have an enhanced probability of survival. Also, the spread in energies of the ionizing electrons is broad so that it is energetically possible to excite high levels of the ions.

5.2.4 Beam Contamination

Accurate measurements require identifying and accounting for the effects of projectiles in the beam other than those of the intended species. In Sections 5.2.1 and 5.2.2 several modes of introduction of spurious species were discussed. All of these modes were associated with collisions of ions in the residual gas along the path from the ion source to the collision chamber, causing spurious species to appear at the output of a mass charge-to-ratio selector. Often, spurious species of this type can be identified by reducing the collision chamber pressure to background level and subjecting the beam leaving the collision chamber to an analysis employing an electrostatic deflector. Such an analyzer separates particles according to energy-to-charge ratio and identifies nearly every type of spurious ion species produced by the gaseous collision processes.

It is always possible that a small fraction of beam contaminants is introduced by collisions of beam particles with solid electrode surfaces or with the edges of slits used to collimate the beam as it passes from the accelerator to the collision chamber. Generally, such species can be identified by an energy-to-charge analysis of the kind discussed above, because particles scattered from solids usually suffer an energy loss, dissociation, or change of charge.

In some cases it may be possible for the beam to eject species of other atoms or molecules from electrode surfaces or to pick up atoms from the electrode surfaces. These effects have received very little attention in collision experiments.

Whenever contaminants can be identified, their contribution to the observed yield of collision products should be determined either from calculations based on known cross sections or by artifically causing a change in the fraction of the contaminant to assess its effect on the measured cross sections.

Contamination particles in the beam that have characteristics similar to the dissociation fragments produced in P-type experiments (such as dissociation fragments produced upstream from the collision chambers), can often amount to a substantial fraction of the observed product ion signal. The usual way of eliminating these is to observe the fragment production as a function of collision chamber pressure, and to use the slope of the curve in determining the cross section.

Electrons can sometimes form a significant beam contaminant in T-type measurements. When an electric field is used to collect slow dissociation fragments of a bombarded gas, care must be exercised to see that secondary electrons produced at the entrance or exit apertures by beam ion impact on solid surfaces are not drawn into the reaction region, thus producing dissociative collisions. Ions of keV energy often have secondary emission coefficients of the order of unity, and it is quite possible to produce electrons at a rate equal to or greater than the beam current if a substantial portion of the beam strikes a surface surrounding an aperture. These problems can usually be eliminated by suitable design provisions.

5.2.5   Measurement of Particle Flux

The beam particle flux is often measured in order to determine a collision cross section. This refers to quantity I_b in Eqs. 3.2 to 3.8. The significant quantity is the flux traversing the collision chamber, and this is usually deduced from beam measurements downstream from the chamber. If too high a pressure is employed in the collision chamber, the emergent beam particle flux may not represent the true flux at other points in the interaction region because of beam attenuation in gaseous collisions. It is always essential to ascertain the exact degree of attenuation from this cause in order to obtain a meaningful measurement. Much beam-gas work is done under conditions such that not more than a part per thousand attenuation occurs.

A variety of methods has been employed for measuring ion beam particle flux, including Faraday cups, plane electrodes with secondary electron supressor grids, secondary emission multipliers, thermal detectors, solid state particle counters, scintillation counters, and proportional counters. All of these generate a signal, either in the form of an electric current whose average value is proportional to the incident ion current, or a sequence of pulses representing individual particle impacts whose occurrence rate is proportional to the incident ion current. When the beam measurement is derived from a current, the proportionality factor relating the signal to the incident ion current must be determined with due consideration to secondary electron emission from the detector surface, backscattering from the detector surface, and collection of spurious currents of electrons which may be produced at apertures.

Secondary emission multipliers and detectors using external electron amplification are always subject to a dependence of the calibration factor on incident ion energy and atomic or molecular species.[95,189] They may also be subject to a dependence of their calibration on the excited state of the impinging particles. Thermal and secondary emission types of detector may exhibit variations of their response with small movements of the impinging beam.

All of the detectors just listed are capable of responding to neutral atoms and molecules, as well as ions. The Faraday cup detector responds to neutrals if a subsidiary electrode is employed to draw secondary electrons away from the surface on which the particles impinge.

Solid state detectors, scintillation detectors, and proportional counters deliver pulses whose amplitudes are approximately proportional to the total ionization or photon yield produced by impinging particles as they come to rest within the sensitive detector volume. The pulse height produced by a single impact is little affected by the excitation state or charge state of a given atomic or molecular species. For these reasons these detectors are greatly preferred to all other types for precise measurements. They are free from most of the difficulties normally encountered in current-type detectors except that of possible count loss by incident particle backscattering. The latter effect may be troublesome at particle energies less than 1 keV. Of the three types of counter detectors mentioned, only the proportional counter is usable for individual particle counting at this low an energy. The solid state and scintillation counters require care to avoid background noise interference at energies lower than around 10 keV for protons. A somewhat higher energy limit applies to heavier particles.

Secondary emission multipliers sometimes have sufficiently low dark current that the pulses produced by individual incident ions or neutral particles can be counted. In this mode the efficiency is equal to the fraction of the incident particles that produce a pulse larger than the acceptance threshold of the pulse counting apparatus. For incident ions of energy less than around 1 keV, the average secondary electron yield per particle may be less than unity, so that even if such a detector is working perfectly within its inherent limitations some loss of counts must occur. Usually the pulse height spread associated with this type of detector is broader than that associated with a proportional counter. Hence identification of particles by pulse height analysis is not easily accomplished with the secondary emission counters.

All detectors that depend on the phenomenon of secondary electron emission are subject to some degree of variability of efficiency due to changes in the state of the sensitive surface. Even single-atom layers can control this process. Experimenters utilizing this type of detector should give careful consideration to the effect of sensitivity changes during the course of a measurement. It is conceivable that the relative electron yields as well as the absolute electron yields of various ion species can change with the degree and type of surface contamination.

5.2.6   Beam Particle Blockage

A potential source of experimental error in cross section measurements is the blockage of a portion of the beam by collision chamber exit aperttures, magnetic pole pieces, deflection plates, detector apertures, and other obstructions in such a way that the beam flux measured by the detector does not represent the total beam flux through the collision chamber. In considering the possibility of such blockage it is important to consider that stray fields can cause the beam particles to follow paths other than the intended ones visualized on the drawing board. It is also very important to recognize that the fringing fields of electrostatic and magnetic analyzers through which beams sometimes pass to the detector can cause deflections normal to the plane of the desired deflection. A good example is to be found where an ion beam enters or leaves a magnetic deflector at oblique incidence (i.e., so that the normal projection of the beam axis on the pole face subtends an angle other than 90° with the edge of the pole face). This situation can either focus or defocus beam particles. Ewald and Hintenberger[5] discuss this focusing effect and provide formulae for calculating the focal length of the equivalent thin lens. When properly exploited this focusing principle can be used to advantage, but when not recognized it can lead to errors in beam measurements.

Some authors have reported that projectile beam current saturation occurs when a rectangular collector slit is broadened beyond a specific amount in one direction without discussing the sufficiency of the aperture dimension in the other orthogonal direction. In general, the beam divergence in two orthogonal directions is not the same, and one must perform tests for the sufficiency of both slit dimensions. Such inequality of divergence is especially likely to occur when the beam is collected after it passes through an electrostatic or magnetic deflector.

5.2.7   Influence of Secondary Collisions

It is important to insure that the fast collision products generated by dissociation or charge exchange of the projectiles do not significantly contribute by gas collisions to the production of the collision products that are to be observed. A test for a departure from linearity of product signal versus collision chamber pressure usually reveals the presence of processes such as these, but it is advisable to perform independent calculations of the secondary particle production from known cross sections whenever possible. It is often feasible to perform cross section measurements for suspected secondary processes with the same apparatus merely by selecting a different beam.

A further source of secondary collision effects is the excitation of beam particles through collisions with gas molecules upstream from the point at which a dissociative collision occurs. The most easily obtained evidence for this effect is a nonlinearity of product signal with pressure.

5.2.8 Angular Divergence and Density Profile

When cross sections which are differential with respect to the emission angle of either a target dissociation fragment or a projectile dissociation fragment are obtained, several additional factors need to be considered concerning the beam specification. This section treats some of these considerations.

The angular resolution of a differential scattering cross section measurement is determined by the angular spread of the beam particles in the interaction region, as well as by the solid angle subtended by the product particle collimator and its subsequent energy or mass filters. The function $\epsilon(\mathbf{a}, \mathbf{x}, \mathbf{s})$ employed in the equations of Chapters 2 and 3 depends on the beam angular spread, as well as the detector geometry. This function must be evaluated as precisely as possible in a differential angular distribution measurement.

In differential angular distribution measurements collimation of the incident beam and product particles is employed to limit the angular spread of the particles involved. Usually the range of scattering angles covered by the detector collimator is a function of the position of the scattering event in the interaction region; the relative weighting of each position in the interaction region is a function of the beam current profile within the collision chamber. The effect of variations in this profile on the measurement needs to be considered. In mathematical terms, the functions $N_b(\mathbf{x})$ and $f_b(\mathbf{x})$ discussed in Chapters 2 and 3 define the variations in beam current density.

Where small changes of direction of beam ions due to stray fields can distort an angular distribution, it is important to eliminate stray fields or to assess the associated errors. It is wise to examine angular distributions for symmetry about the supposed zero scattering angle. Stray fields often produce asymetries in such distributions.

5.2.9 Spontaneous Dissociation

Spontaneous dissociation is discussed in Chapter 3. Consideration must be given to beam particle spontaneous dissociation upstream from the collision chamber as a source of beam contamination. It is also necessary to consider the possible effects of spontaneous dissociation of projectiles between the collision chamber and beam detector as a source of beam attenuation.

5.3   GAS TARGET SPECIFICATION

The majority of collisional dissociation studies have employed the beam-gas method. In order to perform a well-defined cross section measurement using this method, a number of critical factors related to the composition, temperature, and density distribution of the gaseous reactant must be considered. Sections 5.3.1 through 5.3.5 deal with these factors in detail.

5.3.1   Number Density

In order to determine by the beam-gas method a collision cross section which is independent of other cross section data, it is necessary to know the density of the gaseous reactant atoms or molecules throughout the interaction region (see Chapter 3). This is usually done by measuring the density at one point in the reaction chamber and by calculating from kinetic theory the ratio of the density at any other point in the reaction region to that measured at the fixed point. When densities are chosen for the collision chamber so that secondary interactions are negligible (see Section 5.2.7), molecular flow conditions usually prevail throughout the collision chamber, but such conditions do not always hold in regions in front of orifices through which gas is admitted to the collision chamber. It is necessary to ascertain that inlet apertures do not project a dense directional jet either into the pressure sensor or the collision region so as to cause departures from molecular flow conditions.

At the densities corresponding to single collision conditions (usually $<10^{-3}$ torr) it is possible to employ an ionization gage for the density determination at the chosen fixed point in the collision region provided that the gage has been calibrated against a standard with the gas to be used in the collision measurement. Ionization gages have an instantaneous linear response to changes in density and are therefore extremely convenient for the measurement of relative densities. These gages have been widely used for relative density measurements throughout entire measurements, and random and/or systematic drift characteristics have been seldom noted.

The most common means of calibrating ionization gages is the McLeod manometer. Unfortunately, such manometers are subject to a number of sources of error. Two of these have not been recognized in the bulk of the dissociation literature: (1) the effect of pumping in a cold trap placed between the McLeod gage and the gage under calibration, and

(2) the effect of transpiration density gradients caused by temperature gradients along the connecting gas lines between the McLeod gage and the gage under calibration. Both of these effects cause errors whose magnitudes depend on the actual value of the gas density, the diameters and lengths of the connecting tubes, and the temperature profiles present in the system. Corrections are possible in principle when the problems are recognized. Often a large reduction in the magnitudes of the corrections can be made by judicious design. These effects have been summarized and more completely described by Barnett and Gilbody.[4] In that summary it is indicated that errors of the order of 5% can occur for light gases like He, and as large as 40% for heavy gases such as Xe, in typical systems. When these effects are not explicitly noted in experimental procedure writeups, due caution should be used in relying on the authors' error estimates. The pumping effect tends to make the McLeod gage register a lower gas density than that present at the gage being calibrated. An error of this sign leads the investigator to calculate a cross section value higher than the value which would be obtained if the pumping error were properly corrected. The transpiration effect can introduce cross section errors of either positive or negative sign.

Several newer methods of gage calibration are discussed in the Barnett and Gilbody summary.[4] These show definite promise of reducing density measurement errors below that achievable with McLeod gages, but the newer methods have not yet received wide application in cross section measurement.

5.3.2 Number Density Profile

An absolute cross section determination by the beam-gas method requires a knowledge of the function $f_a(\mathbf{x})$ appearing in Eq. 3.1 and defined in the accompanying text. This function can be determined in principle by the use of a movable gas density probe, however, in the entire dissociation literature no such refinement has been used. Variations in density usually exist at entrance and exit apertures of the collision chamber and can be estimated assuming molecular flow conditions. Such conditions exist when the mean free path of the gas molecules is larger than the collision chamber dimensions.

Ordinarily, variations in gas density in directions normal to the beam direction can be ignored because of the smallness of the diameter of beams typically employed.

The detector in a differential cross section measurement views a certain finite portion of the interaction region. In general, the scattered particles coming from each portion have a different mean scattering angle. The weighting given to different portions of that region at a given angular setting of the detector depends on the gas density profile across the region and, of course, the beam density profile as well. In some cases a shift in the gas profile can change the apparent differential cross section. The mathematical description of the gas density profile is expressed by functions $N_a(\mathbf{x})$ and $f_a(\mathbf{x})$ in Chapter 2.

5.3.3   Contamination

The assessment of collision chamber contaminants has received very little attention in the literature. However, with the present commercial availability of many convenient residual gas analyzers it is feasible for most investigators to conduct careful studies of this important error source. Once contamination is estimated, it is necessary to establish the contribution of the contaminating gases to the measured cross section. This can be done by referring to known cross section values, by varying the partial pressure of a contaminant to assess directly its contribution to the cross section, or by a direct measurement in the same apparatus using a pure sample of the contaminant gas.

Residual contamination present in the vacuum system due to evolution from walls and pumps is usually assumed to have a partial pressure independent of the leak rate of the reactant gas into the collision chamber. The validity of this assumption implies that a cross section determination based on the slope of the product yield *versus* gas density gives a correct cross section independent of the background gas. This assumption has rarely, if ever, been tested. A built-in residual gas analyzer to sample the composition of the collision chamber gas as a function of the gas leak rate would allow checks on this assumption.

When contaminant concentrations and their variation with collision chamber pressure are known, corrections to apparent cross sections can readily be made based on known contaminant cross sections.

5.3.4   Thermal Effects*

Assuming that equilibrium thermodynamics holds, the gas molecules or atoms in a gas target at available laboratory temperatures, even up to 3000°K, are predominantly in the ground electronic state. The higher vibrational and rotational states may, however, be substantially populated. The internal states of molecules of a gas target may play a significant role in cross section measurements at low energies, but at the energies considered in this review the effects of target molecule internal energy states probably are not important. A possible exception

* This section is reproduced substantially as written by F. W. Bingham.

is when the gas temperature is high enough to populate vibrational and rotational levels in which the molecules exhibit a considerably different range of internuclear distances from the ground level.

To examine the effect of thermal motion on cross section measurements consider an idealized experiment: particles in a perfectly collimated beam of energy T_b collide with target atoms in a gas, and a detector with perfect collimation receives scattered particles at angle χ, with respect to the beam direction. If the target particles were at rest, all the collisions would involve the same precollision relative velocity (determined by T_b); furthermore, the detector would receive only particles scattered exactly through angle χ. The detector count rate could then be used to calculate a differential cross section for scattering at energy T_b into angle χ.

In a real target gas, however, there is a thermal distribution in the velocities of the target atoms. Accordingly, there is a distribution in the colliding particles' relative velocities before collision. It is convenient to consider this distribution in two parts: a thermal distribution in the magnitudes of the relative velocities, and a random distribution in their directions. Because of the distribution in magnitudes, the detector count rate cannot be used to deduce a cross section characterized by T_b or the projectile velocity; even with perfect beam and detector collimation, the deduced cross section must be an average over all the relative collision speeds. Because of the distribution in directions, the precollision target velocities, are, in general, not parallel to the beam direction; angle χ is therefore not a measure of the true scattering angle for particles that reach the detector.

Whether these two thermal-motion effects significantly influence a cross section measurement depends on the details of the experiment. When the projectile velocity is much greater than the thermal target velocity, it is frequently quite accurate to treat the data as though the target atoms were at rest. Since the measurements reviewed in this report involve bombarding energies greater than 10 eV, few show effects due to target motion. However, the effects may be very pronounced in certain kinds of experiments reviewed here; under some conditions measurements of both total and differential cross sections require correction for target motion. This section presents only a brief review of these conditions.

There are two particular kinds of differential cross section measurements that are especially susceptible to thermal-motion distortions.

1. Experiments that use gas targets at elevated temperatures involve target velocities which may be large enough to make the relative-velocity

effects significant. Confirmation of the accuracy of such experiments requires demonstration that thermal-motion effects are negligible. Only one experiment at elevated gas temperature appears in this review.[6]

2. Measurements of angular distributions of the slow target particles recoiling from collisions generally show strong thermal-motion distortions. Most of these target particles receive so little kinetic energy that their thermal velocities before the collision represent an appreciable fraction of their velocities after the collision. Thus the original random angular distribution of their trajectories contributes a significant spreading to their postcollision angular distribution.

The random direction of target thermal momentum also broadens the energy distribution of the dissociation fragments of target molecules. Chantry and Schulz[190] have shown that the broadening may be surprisingly large. Their analysis indicates, for example, that in one previously reported experiment the energy distribution among nominally 2-eV O^- ions from O_2 dissociation had a width at half maximum of 0.56 eV. Although the Chantry and Schulz paper deals directly with electron-induced dissociations, the analysis is also applicable to reactions induced by heavy projectiles.

A theoretical treatment of the distortion of differential cross sections by thermal motion appears in a paper by Russek,[191] who has calculated, for an arbitrary center-of-mass cross section, the cross section as measured in the laboratory system. Although the general result, expressed as a function of target gas temperature, is very complicated, Russek's work also includes calculations for the special cases of Coulomb and hard-sphere scattering. These results are useful in determining when a measurement may be expected to show thermal-motion effects and how severe the distortions may be.

Thermal target motions do not generally affect measurements of total cross sections which involve collection of all the ions produced by a reaction, regardless of their trajectories. However, when a total cross section is measured near threshold or near sharp structure in the cross section's dependence on collision velocity, thermal motion may influence the result. As mentioned above, there is a distribution in relative velocities before collision; thus there is also distribution in total cross sections for the collisions taking place in the target gas. If these total cross sections vary widely over the distribution in relative velocities, the measured cross section will be difficult to interpret. Of course, target motion does not affect the measurement if the total cross section is independent of collision velocity over the range covered by the vector sum of the projectile and thermal target velocities.

In all the experimental arrangements reviewed in this discussion, target thermal motion distorts the true distributions (for example, in energy or angle) among the particles emerging from atomic collisions. The general problem of removing thermal-motion effects from data is that of unfolding an instrumental distortion from a measured distribution; an observed distribution takes the mathematical form of a convolution integral over the product of the true distribution and a distorting apparatus function.[192] There is, unfortunately, no general solution to this kind of integral equation, although in many cases special techniques can provide accurate approximate solutions.

5.3.5 Influence of Secondary Collisions

The collision chamber pressure must be low enough so that neither beam nor collision product attenuation is significant. A demonstration of the absence of projectile beam attenuation discussed in Section 5.2.5 does not necessarily imply the absence of product attenuation, since in the two cases different collision processes are involved and the thickness of the absorbing gas may be different for the two species. In T-type measurements the dissociation fragments may be low enough in energy so that atom interchange reactions of large cross section occur between the fragments and the target gas molecules. If the dissociation fragments have an accidental charge exchange resonance with the target gas molecules, very severe attentuation may occur.

5.4 PRODUCT SPECIFICATION

The majority of dissociation measurements depends on the observation of an identified dissociation fragment from one of the two reactants. In this section the procedures needed for identification, energy specification, emission angle specification, and flux measurement are discussed.

5.4.1 Atomic and Molecular Species Identification

The identification of dissociation fragments in T-type experiments is usually limited to the ion fragments. These commonly fall in the kinetic energy range of 0 to 10 eV, are emitted in all directions from the interaction region, and consist of several different species. As there is no readily available means of determining the mass or charge of a species separately, the universal practice is to perform a determination of mass-to-charge ratio.

In P-type measurements, the dissociation fragments tend to preserve approximately the direction and velocity of the projectiles undergoing collisional dissociation. At high beam energies, the spread in velocity of the fragments is usually a small percentage of the mean velocity, and it is often possible to perform a mass-to-charge ratio analysis with a much simpler analyzer than in the T-type experiment. An electrostatic or magnetic deflector alone may suffice. At low beam energies, where neither the direction nor velocity of beam molecule dissociation fragments is well-defined, it is necessary to treat the fragments essentially as described for T-type measurements.

When the nature of the molecular reactants is such that two or more distinct atomic species having nearly the same mass-to-charge ratio can be produced as dissociation products, special attention must be given to identification of the products.

5.4.2   Emission Angle and Kinetic Energy Discrimination

The mass-to-charge ratio analysis of dissociation fragments almost always entails a limitation on the accepted emission angle from the collision region or the accepted kinetic energy. Although the angular and energy ranges accepted may be changed by adjustment of the system parameters, a measurement performed with a given set of parameters is always a differential cross section measurement, and it has meaning only insofar as the transmitted particle characteristics are fully described. The mathematical entities for the expression of the transmission of the product analyzer are the functions $\epsilon(\mathbf{x}, \mathbf{s}, \mathbf{a})$ and $F(\mathbf{a}, \mathbf{s})$ introduced in Chapter 2. It is important to specify these functions completely in order to render a measurement well-defined.

Some authors have reported that ion current saturation of dissociation fragments occurs when a collector slit is broadened beyond a specific amount in one direction without discussing the sufficiency of the aperture dimension in the other orthogonal direction. In general, the beam divergence in two orthogonal directions is not the same, and the investigator must perform tests for the sufficiency of both dimensions. Such inequality of divergence is especially likely to occur when the beam is collected after it passes through an electromagnetic or magnetic deflector.

5.4.3   Detector Efficiency

The discussion of detector efficiency in Section 5.2.5 applies to dissociation fragment detection as well as beam detection. Owing to the relatively weak signal produced by dissociation products, compared to beam particles under highly desirable single-collision conditions, secondary emission multipliers have been used for signal enhancement. In such cases the variations in detector efficiency as a function of fragment mass and charge must be made as a prerequisite to placing any confidence in the accuracy of the relative yields of different fragment species or the yield cross sections.

5.4.4   Secondary Collision Effects

A potential source of experimental error is the attenuation of a particular collision product by a gaseous collision while the product is en-route from its point of formation in the interaction region to the detector. This type of collision, if not corrected, reduces the measured product intensity at the detector and leads to an underestimate of the cross section. As the fractional loss is proportional to the pressure of the gas along the path, forced variation of this pressure generally discloses an attenuation effect. It is especially difficult to deal with attenuation in the collision chamber, since variation of that pressure changes the production rate and may vary the primary beam attenuation.

Another type of secondary effect to be considered is the production of spurious secondary ion species by interaction of the collision products with the gas in the collision chamber. These spurious secondaries can be identified by their variation as the second power of the collision chamber pressure.

It is necessary to ascertain that the observed collision products come from the interaction region and not from the walls of the collision region or from ion-molecule reactions in the gas surrounding the beam. The location of product origin can be localized through the use of collimators movable across the interaction region.

5.4.5   Stray Field Effects

When dealing with ion collision products of kinetic energy less than around 10 keV, it is necessary to consider possible deflection of ion trajectories by stray fields. Such fields may occur through electrical charging of surfaces by the beam or by ions or electrons produced in the gas by the beam. In some instances, the space charge of the beam may produce potentials of the order of a few volts between the beam and the collision chamber wall. Such fields can drastically effect the ion trajectories, and electrons or ions may even be entrapped in a cylindrical region around the beam. Stray magnetic fields can also introduce directional deviations of the beam.

5.4.6   Spontaneous Dissociation

The possibility that an observed molecular dissociation fragment is unstable should be examined. If a fragment has a lifetime for further dissociation of the order of the transit time of the specie from the point of formation to the detector, there will be an attenuation of the fragment signal at the detector. A side-effect of this type of attenuation is the production of secondary dissociation products which may be confused with direct collisional dissociation products. This type of attentuation is path-length dependent but not pressure dependent.

CHAPTER 6

QUALITATIVE CONCLUSIONS BASED ON EXPERIMENTAL DATA

6.1 GENERAL CONCLUSIONS

The general impression we have reached from a study of the experimental data on dissociation is that the criteria listed in Chapter 5 were given scant discussion in most of the data-bearing papers. Except for a few special cases, each of the criteria has an important place in the discussion of a measurement because it relates to the definition of the measured quantity or because it designates a source of experimental uncertainty deserving explicit mention and evaluation. Based on general inattention to the criteria in reports on the measurements, we conclude that the bulk of the experimental data is not well-defined in a quantitative sense. We further conclude that the uncertainty limits stated in the papers are rarely supported by adequate discussion. It is unusual to find even a listing of the error sources considered by the investigator.

In the discussion below, frequent reference is made to P- and T-type measurements. These designations refer, respectively, to projectile molecule dissociation measurements and target molecule dissociation measurements as defined in Chapter 4.

Criteria often violated in P-type measurements are that the excited-state content of the beam is unknown and that the precise definition of the dissociation fragment emission angle acceptance limits is not clearly stated. In T-type measurements, criteria often violated are that the dissociation fragment energy and emission angle acceptance limits are not clearly stated. With the exception of T experiments involving H^+ or He^{2+} projectiles, the initial states of the projectiles are not precisely known. Detector efficiency is often inadequately examined in both types of measurement. Both P- and T-type measurements were frequently conducted with McLeod gage calibration of the pressure-measuring

121

instruments, and there is little or no discussion of the pumping and transpiration errors mentioned in Chapter 5. A large fraction of the data was obtained before these error sources were recognized.

In spite of the unsatisfactory status of most of the data, as viewed from the standpoint of our criteria, a number of conclusions of a qualitative nature seem to be well-founded. These conclusions are discussed in the remainder of this chapter. Most of the conclusions refer to H_2^+ dissociation. This molecule is, by far, the most widely studied in regard to quantity of data, energy range, and aspects of dissociation.

An important point to bear in mind in reviewing any experimental data on dissociation is that a given dissociation fragment may arise from more than one fundamental process. In addition to originating from several fragmentation modes (see Section 6.4), an observed fragment can be produced in more than one excited state, and the unobserved fragments accompanying an observed fragment can also be produced in several excited states. These facts often complicate comparison of theory and experiment.

6.2 ION SOURCE EFFECTS

Tunitskii et al.,[7] in a study of H_2^+ dissociation, obtained the first evidence of a dependence of the cross section for collisional dissociation of a molecular ion on the kinetic energy of the electrons used to produce the ions. This effect was also observed by Kupriyanov et al. in a number of experiments (see refs. 8, 22, 58, 62, 102, 125, and 126) involving H_2^+, D_2^+, CH_4^+, $C_2H_3^+$, CH^+, CH_2^+, and $C_2H_2^+$ projectiles. McGowan and Kerwin[9,11] observed similar effects in O_2^+ and N_2^+, and confirmed the earlier evidence[7] of an effect on H_2^+ ions.[10] These studies leave no doubt that it is essential to specify the energy of the electrons employed to produce the ions in order to completely define a P-type measurement. The origin of the cross section dependence on ion source electron energy is undoubtedly that the ions are produced in a varying population of excited electronic or vibrational states, that the population depends on electron energy, and that the dissociation cross section depends on the population.

McGowan and Kerwin[9,11] have associated changes in the N_2^+ and O_2^+ dissociation cross sections with the onset of several successive electronic excited states with increasing electron energy. The deduction of cross sections for ions in the various states depends on knowing both the relative electron impact production cross sections for each excited state over the whole energy range of the electrons employed, and to what extent the excited-state populations are altered by collisions in the ion source and by radiative transitions while the ions are in transit to the collision chamber.

Riviere and Sweetman[47] compared the dissociation cross sections for H_2^+ ions produced by two methods: (1) directly in an ion source, and (2) from the collisional dissociation of H_3^+ ions. The experiments were performed in the energy range 280 to 670 keV with H_2 as the target gas. For the dissociation mode $H_2^+ \rightarrow H^+ + H$, the ions from method 2 showed a cross section $7 \pm 4\%$ higher than those from method 1. For the dissociation mode $H_2^+ \rightarrow H^+ + H^+$, the difference in the cross sections for ions from the two methods was smaller.

Barnett and Ray[28] performed an experiment similar to the study above in the energy range 50 to 200 keV. They found that the dissociation cross sections for H_2^+ ions produced by method 2 were from 1.2 to 1.4 times higher than those produced by method 1. The larger ratio was observed at the higher energy.

Kupriyanov and Perov[101] have shown that the cross section for H_2^+ dissociation in collision with Ne at 3.8 keV depends strongly on the gas from which the H_2^+ was formed in an electron-impact ion source. The gases H_2, C_2H_4, n-C_4H_{10}, and $C_{12}H_{24}$ were studied. Some evidence was obtained that H_2 was present in the ion source with the polyatomic molecules due to thermal dissociation of the polyatomic molecules on hot surfaces. Corrections were not made for the direct production of H_2^+ from thermally produced H_2, but a factor of 2.4 variation in the cross section was observed among the gases studied.

In a related study, Kuprianov and Perov[102] observed a strong dependence of the cross section on H_2^+ dissociation in collision with Ne when the H_2^+ was produced by variable energy electron bombardment of CH_4. Analogous results were also obtained for D_2^+ from CD_4. These studies indicate that the electron bombardment energy affects the population of internal states of dissociation fragments produced in the ion source.

Lindholm et al. (see refs. 130, 136, 139, 151, 153, 164, 179, 182, 183, and 185), Petterson,[184] von Koch,[170] and Sjogren[180] have made a number of measurements of the T type in which projectile ions of one species were formed by electron impact on more than one ion source gas. Many of these experiments show a definite effect of ion source gas on the *target* molecule fragmentation pattern and relative fragment production cross section, suggesting that the ions in question were formed in varying populations of excited states as the gas was altered, and that the internal state of a projectile atom influences the cross section for the dissociation of the struck molecule.

Barnett and Ray[28] have compared the cross sections for dissociation of H_2^+ ions produced in a radio-frequency ion source and in a Phillips ion-gage source. They found an overall random spread of about $\pm 5\%$ in the data points, but discovered no significant difference in the cross sections obtained from the two methods, nor any change in the cross sections with variation of the parameters of the sources. These results were obtained throughout the H_2^+ ion kinetic energy range of 40 to 200 keV.

Cross sections were also found in the study above to be insensitive to a change from a mechanical to a palladium metal diffusion leak in the ion source gas feed system. Since hydrogen diffuses through the palladium lattice in the atomic state, it was considered that some of the gas emergent from the surface might be atomic and that the gas reaching the collision chamber might comprise some atoms. The result indicated that if any atomic hydrogen were present, it made no significant change in the apparent cross section.

McClure[27] observed that the cross section for the collisional dissociation of 10-keV H_2^+ ions in collision with H_2 to form protons or H atoms was dependent on the kinetic energy with which the H_2^+ ions emerged from a Penning discharge ion source. With the source he used, the emergent ions had a spread of energy of several hundred eV. In the same experiments, McClure observed a 20 to 30% dependence on the ion source pressure of the cross section for dissociation to yield a proton.

Williams and Dunbar[25] observed substantial effects on the cross section for H_2^+ dissociation on gas pressure and magnetic field in a radio-frequency ion source. The size of the effect was largest (30 to 50%) at low energies (6 keV). At the higher energies investigated (50 keV), where the energy range overlapped that of the ion source studies of Barnett and Ray,[28] the effect observed by Williams and Dunbar was small.

Chambers[109] presented evidence of a dependence of the cross sections for H_2^+ and H_3^+ ion dissociation on the strength of the magnetic field applied to a radio-frequency ion source used to produce the ions. For H_3^+ ions, variations of a factor of 4 in the cross section were observed, whereas for H_2^+ ions, variations of the order of 20% were observed.

Caudano and Delfosse[107] have determined the energy distributions of forward-directed H^+ dissociation fragments generated in $H_2^+ + M$ collisions at 10 keV, where M is H_2 or Xe. They found a very strong dependence of these distributions on the voltage applied to the Penning source which produced the H_2^+ ions.

Some dissociation measurements in which the states of both reactants are well-defined are the measurements of T type, in which the projectiles were H^+ or He^{2+}. Perhaps a few other cases exist in which the projectiles were singly or doubly ionized heavier atoms and in which the ion source electron energy was below threshold for excitation of the selected ion. We have not made a systematic survey in search of such cases. In all measurements of T type, the target gas molecules are in a thermal population of rotational states appropriate to the target gas temperature. Vibrational or electronic excitation is probably negligible at the temperatures employed in existing studies.

6.3 DISSOCIATION FRAGMENT ENERGY AND ANGLE DISTRIBUTIONS

There is a considerable amount of information on the energy and angular distribution of dissociation fragments produced in P-type experiments. In collisions at keV kinetic energies, the fragments typically spread outward from the beam direction by angles of the order of a degree, owing to the mutual repulsion of dissociation fragments. In the case of H_2^+ dissociation, the measurements of McClure[63] show an increase in the angular spread of both H^+ and H dissociation fragments as the H_2^+ energy is reduced from 80 to 5 keV. McClure has shown that the distribution of the H dissociation fragment velocity component, normal to the beam direction, is independent of H_2^+ energy, but that the H^+ fragments have a velocity distribution that becomes broader as the H_2^+ energy increases. The H^+ distributions obtained at 10 keV were compared with angular distributions calculated on the Born approximation, on the assumption that dissociation occurred solely by excitation of the $2p\sigma_u$ state. Fairly good agreement was obtained.

Very detailed energy and angular distributions of H^+ dissociation fragments from H_2^+ have been measured by Gibson et al.[40-43] These have been translated to the center-of-mass coordinate system of the H_2^+ ion after the collision, with an assumption about the inelastic energy loss and scattering of the center-of-mass of the dissociating projectile.

These studies show very interesting structure, including peaks at 90° and definite evidence of anisotropy. The velocity distributions are roughly consistent with the assumption that dissociation occurs via the lowest $2p\sigma_u$ excited electronic state, as predicted by calculations of Green and Peek presented in ref. 63.

Vogler and Seibt[44] have conducted investigations similar to those described above. Very recently Durup, Fournier, and Dong[39] have obtained H^+ fragment energy distributions near a scattering angle of zero degrees in the lab system under conditions of controlled ion-source electron energy. The distributions are profoundly modified by changing the ion source electron energy. The electron energy dependence was qualitatively interpreted in terms of the Franck Condon principle and the available vibrational states of the H_2^+ ion as a function of the electron energy excess over threshold.

Rourke, Shefield, and Davis[31] have examined the energy distribution of fragments from a number of heavy molecule ions containing between two and five atoms. They found that the fragment energies, relative to the projectile center-of-mass system, fell in the range of 0.7 eV for $SF^+ \rightarrow F^+$ to 7.4 eV for $N_2^{+2} \rightarrow N^+$. When the relative velocity of the dissociation fragments is oriented at right angles to the beam direction, substantial angular divergence may occur in the fragment beam. It is therefore clear that very close attention must be paid to angular distributions in $P1$-type experiments in order to collect all the fragments. Whether or not complete collection is demonstrated, it is important to specify the maximum angle at which collection occurred without blockage and, even better, to specify the resolution function $F(\mathbf{a}, \mathbf{s})$ discussed in Chapter 2.

In general, Rourke et al.[31] found triple-peaked energy distributions. The central peak was associated with fragments having very small kinetic energy in the center-of-mass system of the molecule, and the side peaks corresponded to ions having forward or backward velocities relative to the mass center.

For several of the molecules studied by Rourke et al.[31] it was found that the energy spread of the dissociation fragments in the laboratory coordinate system in the forward direction was proportional to the square root of the projectile molecule kinetic energy. This observation supports the assumption that the relative velocity with which the fragments separated was independent of the collision energy.

We quote directly from the discussion in the paper of Rourke et al.[31] the following general conclusions:

"1. Many dissociation reactions are not simple; that is, there is more than one mode of dissociation. The presence of three peaks in the velocity spectrum of a single product ion would imply that there are at least two separate dissociation reactions leading to this ion.

"2. All the product ions that were examined had a single or central peak corresponding to the same velocity as the parent ion. It would thus appear that a large majority, if not all, dissociation reactions can occur with no kinetic energy being imparted to the products. Probably the simplest mechanism for explaining this fact would be a transformation of a very small fraction of the available kinetic energy into vibrational energy sufficient to dissociate the ion.

"3. The side peaks of the two singly charged product ions from the dissociation of doubly charged diatomic ions have the same energy separation and relative intensity. It would thus appear that they result from the same dissociation. The mechanism for this dissociation is probably that involving either (1) an upward transition of the ion to a repulsive state of the doubly charged ion and subsequent dissociation, or (2) an increase in vibrational energy sufficient to overcome the potential barrier leading to dissociation. The latter mechanism depends upon a potential barrier separating the ground state of the doubly charged ion from the two separated ions and having a height above both. The difference between the height of this barrier and the sum of the potential energies of the two separated ions would then be the kinetic energy of the dissociation. CO^{+2} has an appearance potential some 7 eV above the sum of the ionization energies of C^+ and O^+ and thus should have such a potential barrier. The observed kinetic energy of dissociation should be at least 7 eV while the measured value was 5.9 eV. This amount of disagreement could be due merely to errors in measurement. The presence of such a potential barrier in most doubly charged diatomic ions would explain the fact that doubly charged ions are more likely to yield side peaks upon dissociation than singly charged ones."

The energy distribution of C^+ and O^+ fragments produced in the collision-induced and spontaneous dissociation of CO^{2+} ions were studied by Kupriyanov.[50] The collision-induced energy spectra showed dissociation fragments having both high and low kinetic energy relative to the center of mass of the dissociating molecule. The spontaneous dissociation fragments showed a high-energy group of fragments, but the low-energy group was almost absent.

McGowan and Kerwin[9-12] have obtained information on the relative contribution of various specific excited states of N_2^+, O_2^+, and H_2^+ ions on the collisional dissociation of these species by studying relative cross sections as a function of ion-source electron energy in an Aston band apparatus. It was shown that those fragments which were formed with kinetic energy relative to the center of mass of their parent molecules depended in a different way on the excited state of the parent than did those fragments formed without kinetic energy. The dissociation cross sections determined with this apparatus were characterized as lower limits because discrimination against dissociation fragments was thought to exceed that against primary ions.

Valckx and Verveer[93] have shown that the energy distribution of H^+ dissociation fragments from H_2 projectiles incident on H_2 at 70 keV is strongly affected by the magnitude of the pressure of gas in the collision chamber. The range of target-gas thickness investigated was 0.5 to 2.5×10^{15} molecules cm^{-2}. A possible theoretical interpretation is that vibrational population of the H_2^+ ions was altered by collisions which occurred prior to the dissociative collisions at the higher pressures. Similar results have been obtained by Caudano[104] for 10-keV H_2^+ ions incident on H_2 gas in the thickness range of 0.7 to 6×10^{15} molecules cm^{-2}. In both of these studies, a substantial conversion of H_2^+ to H_2 was occurring at the higher pressures according to the cross sections of McClure[27] for the process $\underline{H}_2^+ + H_2 \rightarrow \underline{H}_2$. (Throughout this chapter the underlined symbols represent the same particle or a modified form thereof before and after the collision.) The contribution to the observed H^+ particles from subsequent interactions of H_2 with H_2 is probably substantial, as evidenced by the cross sections measured by McClure[65] for the process $\underline{H}_2 + H_2 \rightarrow \underline{H}^+$.

Investigations of H_2^+ dissociation in collisions with Ar conducted by Vogler and Seibt[44] also show gas thickness effects on the H^+ energy and angular distributions at somewhat lower gas thicknesses.

Ziemba and co-workers[193] observed that, when 5 to 300 keV H_2^+ ions are incident on H_2 or He, the charged ions scattered at 5° from the beam direction consisted solely of H^+ ions. The data indicated that no H_2^+ ion was scattered through this large an angle without dissociating. When H_3^+ ions were substituted for H_2^+, the ions scattered at 5° were H^+ only; no H_3^+ or H_2^+ were found. The ratio of $H°$ to H^+ at this angle was found to be ~0.5 to 0.7 for both H_2^+ and H_3^+ projectiles. Fedorenko and co-workers[24] obtained results analogous to the above mentioned results for H_2^+ scattering in Ar at 24 keV over a larger angular range.

6.4 FRAGMENTATION MODE STUDIES

Among the most informative measurements in the field of collisional dissociation are those in which the partitioning among various possible fragmentation modes is established. Thus far, only H_2^+ dissociation has been studied by these techniques. Guidini[48] and Sweetman[21,46] have used similar techniques to separate the modes $H_2^+ \rightarrow H^+ + H$ and $H_2^+ \rightarrow H^+ + H^+$ in the 30- to 200-keV energy range. It has been found that each mode has a distinct form of cross section energy dependence and that the cross sections for the modes $H_2^+ \to H + H$ and $H_2^+ \to H_2$ which involve charge transfer fall off very rapidly in comparison with the other processes at high energies. Similar behavior is observed with charge transfer of atoms in a similar range of projectile velocities.

6.5 GENERAL PHENOMENOLOGY

The simplest heavy-particle induced dissociative collision investigated thus far is the process $\underline{H}_2^+ + H \to \underline{H}^+$, investigated by McClure.[6] This process was found to yield the same angular distribution of H^+ fragments as the corresponding event produced by an H_2 target molecule at 10 keV. The energy dependence of the cross section, however, was found to be quite different for H_2 and H. The H_2 molecule was not equivalent to two separate H atoms in regard to its cross section for inducing the dissociation of H_2^+ projectiles.

Kupriyanov[149] has concluded, from a study of dissociation of CH_4^+, CH_3^+, and CH_2^+ ions of 1.3- to 3.0-keV energy in collisions with He, Ar, H_2, air, and CH_4, that the *relative* yields of the dissociation fragments are insensitive to the neutral gas. He notes that this evidence supports the idea that the postcollision-state population of the projectiles in states which lead to dissociation is little influenced by the gas giving rise to the excitation.

Lindholm and several other investigators have observed that the fragmentation patterns of a variety of molecular gases are strongly dependent on the projectile ion species inducing the gas molecule breakup for projectile kinetic energies in the range of 5 to 900 eV (see refs. 71, 130, 136, 139, 144, 151–154, 158, 164, 170, 172, 176, 179, and 180–185). This behavior suggests that, at low projectile energies, charge transfer plays a dominant role in dissociation. Some of the data indicate that the excess energy made available when an electron transfers from the neutral molecule to a more tightly bound state of the projectile ion (i.e., the difference in binding energies) is available for internal excitation of the molecule. Reference 183 and those that follow it in the list above present graphical plots of the percentage of each dissociation fragment versus the ionization potential of the projectile which induced dissociation. These plots show drastic dependencies of the fragmentation patterns on projectile ionization potential. It is probable that the apparatus used for this work discriminated against dissociation fragments which were formed with kinetic energy. The discrimination was probably greater for fragments with larger kinetic energy and would tend to be stronger for fragments having velocities parallel to the beam direction.

Kuprianov has deduced[194] an interesting property of the dissociation fragment yields which he observed for the sequence of projectile ions CH_4^+, CH_3^+, CH_2^+, and CH^+.[149] He finds that the mass spectra of the carbon-containing dissociation fragments of the lighter projectiles have the same form as the spectra of the corresponding fragments of the heavier projectiles. Kuprianov has noted that this behavior is consistent with the idea that the dissociative breakup of all of the species in the series occurs by stepwise ejection of a sequence of single hydrogen atoms, and that the residual heavy particle in any stage of the sequence is statistically in the same distribution of states of excitation regardless of the number of preceding steps. The theory of polyatomic molecule dissociation has not yet provided a basis for the understanding of this model, but the model may provide an important clue to the dissociative process in other complicated systems analogous to this series.

Collins[160] has studied C_2H_4 dissociation fragmentation under proton impact at 50 and 100 keV and has compared his results with those of other workers[163,177,178] at higher and lower energies. It appears that substantial changes in the fragmentation pattern occur for proton energies in the range of 10 to 400 keV, but that asymptotic behavior is reached at energies higher than 400 keV. It was pointed out by Collins[160] that Schuler's data[162] do not agree with those of Wexler,[163] used in this composite data correlation.[160] Collins proposes that this discrepancy may be due to unlike fragment discriminations operative in the Wexler and Schuler apparatus.

6.6   STRUCTURE OF CROSS SECTIONS VERSUS ENERGY CURVES

In regard to the variation of collisional dissociation cross section with energy, the behavior of H_2^+ ions is by far the most thoroughly investigated. The cross section to yield a proton fragment apparently decreases roughly in proportion to the reciprocal of the H_2^+ kinetic energy when the energy is greater than 200 keV. In the energy range between 1 and 200 keV, the cross section is relatively constant but has some weak structure. The data for $\underline{H}_2^+ + H_2 \rightarrow \underline{H}^+$ show a double hump with maxima at 10 and 100 keV.[25,100] The results of Guidini[48] for the dissociation mode $\underline{H}_2^+ \rightarrow \underline{H}^+ + \underline{H}^+$ have a maximum at 100 keV with accounts for one of these humps. The cross section for the mode $\underline{H}_2^+ \rightarrow \underline{H}^+ + \underline{H}$ is steadily decreasing with increasing energy in the range of 30 to 250 keV. The cross section for the process $\underline{H}_2{}^+ \to \underline{H} + \underline{H}$ must be dominant over both of the processes above in the energy range of 30 to 100 keV because the yield of H atoms exceeds the yield of protons.[63] This latter process reaches a maximum at 20 keV, based on an analysis of the data of McClure[27] and Guidini[48] presented in ref. 63. This analysis assumed a monotonically increasing cross section for the process $\underline{H}_2{}^+ \to \underline{H}^+ + \underline{H}^+$ at energies lower than 30 keV.

At energies below around 70 keV, substantial variations in cross sections are to be expected from ion source conditions as discussed in Section 6.2 of this chapter. No definitive cross section values can be obtained for this energy region until techniques are developed for producing $H_2{}^+$ ions in well-defined states. There is a possibility of populating only the lowest variation state of $H_2{}^+$ by bombarding H_2 with electrons of sufficient energy to yield $H_2{}^+(\nu = 0)$, but insufficient energy to yield $H_2{}^+(\nu \geq 1)$. This was suggested by Barnett and Ray,[28] but accomplishment has not yet been reported. Selective population of an individual vibrational state is a goal for the future; until state selection techniques are developed for molecule ions, P-type studies can be expected to yield only order-of-magnitude data for low-energy dissociation cross sections.

The methods of Chupka et al.[195] seem to have great potential for selective state population.

CHAPTER 7

THEORY OF DISSOCIATIVE COLLISIONS BETWEEN HEAVY PARTICLES

7.1 INTRODUCTION

Our purpose in this chapter is to review the theoretical literature concerned with collisions between heavy particles that result in the dissociation of one of the collision partners. Obviously molecular systems form the main focal point of interest. In restricting attention to collisions between heavy particles we imply that dissociative processes caused by electrons or photons will be excluded except when the study of such collisions can be utilized to understand interactions between heavy particles. The relative collision energy is restricted to be above several electron volts, hence collisions at thermal or near thermal energies are excluded. The theory of multiple collisions and the theory of collisions involving the exchange of heavy particles, reactive collisions, are not considered either. Many of the dissociation mechanisms of interest to this chapter are discussed in Chapter 2. It is hoped that the literature citations are complete through January 1969; some references through January 1970 are also included. The survey of 1969 should not be considered complete since this review was compiled during 1969 and some of the 1969 material was available only in preprint form.

One additional point must be made to insure that no misunderstanding arises as to our definition of a theoretical paper. A theoretical paper may contain a development associated with the construction or solution of a mathematical model of some physical process. Such a paper is usually abstract in nature and some references cited in this chapter are solely addressed to this aspect of dissociative collisions. A theoretical paper may also be concerned with providing observable quantities as predicted by an assumed mathematical model. The emphasis here is on the latter type of theoretical study. There is yet another type of published information which consists of qualitative discussions proposing mechanisms to explain experimental data that is sometimes equated with a theoretical investigation. Such a discussion receives little or no attention in this chapter unless it goes beyond phenomenology and presents details based on a clearly defined mathematical model.

It is obvious from the comparison of the material presented in Chapter 4 with Table 7.2 that available theoretical results in no way equal the experimental data in terms of the complexity or number of systems studied. Nevertheless, the theoretical studies add much to the understanding of the rather complicated system under consideration in that many comparisons between theory and experiment have been possible. It also appears that work in this area has belatedly passed through a phase of development roughly comparable to that achieved in the area of electron-atom scattering during the decade of the 1930s. (This delay in understanding of collisions involving molecular systems is obviously related to the availability of fairly large-scale computing equipment.) For these reasons, it seems that a review of the literature in this area is appropriate despite the relatively small number of references that can be discussed.

Once a theorist has decided to study a particular system his first obligation is to choose the general theoretical formalism that will satisfactorily describe the problem, and then to incorporate as much available information about his system into this formalism as is practical. As indicated above, the area under review has been preceded by a great deal of study devoted to the simpler electron-atom scattering system with the result that there is little question about the applicability of quantum mechanics, and often classical mechanics can be used. Since we are dealing with the collisional dissociation of a molecule, it is imperative that the structure of the molecule be described in enough detail to define clearly the various events leading to dissociation. The structure of the heavy particle constituting the molecule's collision partner must also be explained in similar detail. The structure of simple molecular systems, that is, ones having few nuclei and electrons, poses a challenge to the theorist; hence it is no surprise that only the dissociation of diatomic systems on collision with the simplest of heavy particles has received attention. As the structure of the target or projectile alone is a difficult problem, it follows that a detailed description of the interaction during collision is a forbidding task. Indeed, this latter problem is a subject of much concern today even for the problem of electron scattering by an atomic target having only one or two bound electrons. The situation dictates that the subject under review will add little to the abstract understanding of the scattering interaction and will be mainly concerned with the application of the simplest of scattering theories. It can be anticipated, however, that the material under consideration contains some novel elements from the theoretical point of view.

Section 7.2 contains a discussion of several inelastic processes that can lead to dissociation and introduces the notation which is used in the subsequent sections. Section 7.3 discusses, more or less in outline form, the various theories and approximations that have been used. Only classical theories and the first Born approximation are considered in any detail. These two frameworks provide the basis for almost all published theoretical studies. Section 7.4 contains our review of the theoretical literature. The main contents of this section are Tables 7.2 and 7.3. Reference to Table 7.2 provides an index to the systems that have been studied and the papers concerned with that particular system. Table 7.3 provides an index to the contents of a given paper, a code defining the theoretical approach, and some amplifying remarks. Section 7.5 contains an informal, and undoubtedly incomplete, resumé of the status of scattering theory as applied to dissociative collisions.

7.2   NOTATION

In this section we briefly discuss the mechanisms that can lead to dissociation and define the notation that is used to describe the content of papers pertaining to dissociative collisions. We have not encountered a theoretical treatment concerned with the heavy particle dissociation of a polyatomic molecule. To reflect this fact the notation is specialized to the case of a diatomic system.

Considering the diatomic system AB, it is necessary to describe its configuration before and after the collision. This is done by adding, in parentheses, a symbol describing the electronic configuration followed by a semicolon and a symbol, or symbols describing the rotational-vibrational degrees of freedom. The symbols are defined in Table 7.1. The same symbols are used to describe the molecule's collision partner T. It is not necessary to specify any internal degrees of freedom for a bare charge T, a single symbol describing the electronic state appears in parentheses for an atomic T, and the convention introduced for AB is used in case T represents a molecule.

The cross section for the collision

$$AB(O; \nu J) + T(O) = AB(N; \Sigma) + T(O) \tag{7.1}$$

is interpreted in terms of the above conventions and Table 7.1 as follows:

TABLE 7.1
The definitions of various symbols used in Chapter 7

Symbol	Definition
O	Lowest or ground state.
ν	Vibrational quantum number.
J	Rotational angular momentum.
ω	A quantum number describing a discrete state of nuclear motion; usually a combination of ν and J.
Σ	A sum over all possible states associated with a particular degree(s) of freedom.
Σ'	The sum Σ with the initial or ground state omitted.
Σ''	A sum that excludes the two lowest states.
$\Sigma''_\sigma, \Sigma'_\pi$	Sums over states with the indicated symmetry. In this case, the prime(s) refers to the omission of the lowest state(s) having the symmetry indicated by the subscript.
I	All possible modes of ionization.
V	All possible modes of vibrational dissociation.
D	All possible modes of electronic excitations (does not include ionization).
R	Indicates that the calculation was done for a fixed value of the magnitude of the internuclear vector $\mathbf{R}$. The magnitude is given in atomic units.
$\bar{R}$	A certain weighted average over a range of R; reference to the original paper is necessary for details.
κ	The cross section is differential in direction and magnitude of dissociation momentum.
$\hat{\kappa}$	The cross section is differential in the direction of dissociation momentum.
$B1$	The first Born approximation.
C	A classical theory as described in Section 7.3.
G	The adaptation of classical mechanics proposed by Gryzinski.
Ru	The use of classical scattering theory based on the Rutherford formula. See Eq. 7.5 and subsequent discussion.

initially AB is in its ground electronic state with rotational-vibrational modes given by J and ν, respectively. Initially T, which is an atom in this case, is in its ground electronic state. After the collision, AB is in the electronic state N, and the symbol Σ appearing after the semicolon indicates that all modes of internuclear motion in the electronic state N are included in the cross section. The atom T is in its ground electronic state after the collision. By reference to Table 7.1 it can be seen that if we add the cross section for reaction 7.1 to the cross section for

$$AB(O; \nu J) + T(O) = AB(N; \Sigma) + T(\Sigma') \tag{7.2}$$

the sum is for exciting all possible states of the target, while exciting all rotational-vibrational modes in the final electronic state N of AB.

Another version of Eqs. 7.1 and 7.2 must be introduced to define the collisions for which some theoretical study is available. The symbols for the rotational-vibrational degrees of freedom are sometimes replaced by a number R. The appearance of this number refers to the situation in which the internuclear separation R is assumed fixed at the indicated value, thus removing the vibrational degree of freedom from the scattering problem. An example is

$$AB(O; 2.0) + T = AB(N; 2.0) + T \qquad (7.3)$$

This refers to a total cross section for AB initially in its ground electronic state with an internuclear separation $R = 2.0a_o$ colliding with a structureless charged particle T. The collision products are AB in the final electronic state, N, with $R = 2.0a_o$ and T.

Pictorially the collision between AB and T takes place with a fixed value for the internuclear vector $\mathbf{R}$. The cross section depending on $\mathbf{R}$ is then averaged over all orientations of $\mathbf{R}$, assuming there is no preferential orientation of $\mathbf{R}$. Since typical rotational and vibrational periods are long compared to an estimate of the collision time, except for extremely small collision energies, this view of the collision seems reasonable. The quantum mechanical equivalent to this description, as it occurs in first Born theory, is introduced in Section 7.3 as Approximation 3.

It is necessary to establish the cross sections defined by Eqs. 7.1–7.3 as dissociation events. The occurrence of Σ in Eqs. 7.1 and 7.2 implies that all rotational-vibrational modes in the final electronic state N are to be included in the cross section. Usually some of these states can be expected to be discrete, that is, they do not correspond to dissociation of AB. The way in which the dissociation events are isolated from this cross section is discussed in Section 7.3, especially in Approximations 4, 5, and 6. As for Eq. 7.3, if the value of the potential curve for the electronic state N at R is above the dissociation limit of N, it is presumed that dissociation of AB will occur.

Yet another example of the notation is

$$AB(O; R) + T(O) = AB(N; \kappa) + T(O) \qquad (7.4)$$

which refers to a cross section differential with respect to the variable κ. The quantity κ describes the direction and magnitude of the dissociation momentum of AB, usually referred to the dissociation limit of AB in the electronic state N, after the collision has taken place. The cross section defined by Eq. 7.4 and the cross section differential in either the direction

$\hat{\kappa}$ or the magnitude of the dissociation momentum $|\kappa|$ associated with the fragments of AB are the only differential cross sections encountered in this review.

The notation introduced here is considered in somewhat more detail in the following section. The formulae for the various cross sections are introduced and the meaning of the symbolism with respect to a quantum mechanical or classical description is developed.

This somewhat elaborate notation is necessary to describe the various theoretical cross sections pertaining to dissociation events. Obviously parameters, or eigenvalues, for the internal degrees of freedom of the molecule and its collision partner must be defined both before and after the collision. Also, although it is possible to measure or calculate the cross section for a given set of eigenvalues, it is rarely done. Experiment usually measures contributions from a large number of events and, if one wishes to compare theory with experiment, it is necessary to include all these events in the theoretical prediction. Hence the symbols Σ, Σ', I, V, and so on, are introduced to define the types of sums encountered in the theoretical studies.

7.3 THEORETICAL TECHNIQUES

As pointed out in the preceding sections, only the simpler approximations have been used to discuss dissociative collisions. Specifically, adaptions of the cross section for scattering of two bodies interacting through a static coulomb potential and the first Born approximation form the basis for most of the work that is discussed. These theories are well documented but, since the description of dissociation events within the framework of these theories has some novel features, a resumé is given of the various equations that have been used. Other theoretical approaches, such as the impact parameter method and the sudden approximation, have also been used to study dissociation. Only a few studies utilizing these methods are available, so their application to dissociative collisions is not discussed in this section. However, pertinent results based on these theories are included in subsequent sections.

First, the application of two-body scattering theory to this manybody problem is described, and this discussion leads us to classify the two-body approach as a classical theory. Next, the first Born theory is described. All available applications of the first Born theory involve additional approximations. These additional approximations are then discussed and each is given a number to facilitate the subsequent description of the literature.

Two-body scattering theory was first applied to the type of dissociative collision under consideration by Salpeter[196] in 1950. This binary collision model is classified here as a classical theory; the implication that all classical theories are based on the binary model is not intended. In the binary approach, one focuses attention on one particle α in the molecule, having charge $Z_\alpha e$, and one particle a in the collision partner, which has charge $Z_a e$. The magnitude of the electron charge is defined as e, and Z_i is the charge number, including sign, of the ith particle. It is assumed that all remaining particles making up the collision partners are passive spectators during the collision, and that the particle α is stationary with respect to the center of mass of the molecule and a is stationary with respect to the center of mass of the molecule's collision partner. A static coulomb interaction between particles a and α is assumed. The cross section for this binary scattering process is

$$\frac{dQ}{d\Omega} = I(\theta) \tag{7.5}$$

where the scattering amplitude $I(\theta)$ as a function of scattering angle θ is given correctly by the Rutherford formula, by the first Born approximation, and by the exact solution to the scattering problem.[197] Here Q is the total cross section and $d\Omega$ is the element of solid angle for the scattered particle. Particle α must undergo some inelastic transition which has a known energy threshold ΔE_α. It is further assumed that the probability for this inelastic event is unity if the kinematics of the collision allows an energy transfer greater than or equal to ΔE_α. Multiplying $dQ/d\Omega$ by this probability function, using the Rutherford formula $I(\theta)$, and integrating Eq. 7.5, one arrives at a cross section for an inelastic event with energy transfer greater than ΔE_α. It is usually assumed that the scattering angle θ is small, and an approximation like $\sin\theta/2 \sim \theta/2$ is made before the integration over θ is carried out. The total cross section thus obtained is inversely proportional to the collision energy and the minimum excitation energy ΔE_α and directly proportional to $(Z_a Z_\alpha)^2$. If the molecule's collision partner is assumed to be an atom with Z_a electrons, each of which contributes to the excitation of the particle α, then the sum of these contributions gives an additional term proportional to $Z_a Z_\alpha^2$. It should be pointed out that the cross section proportional to $Z_a Z_\alpha^2$ can be interpreted as including events in which both the particle α and the electrons associated with the molecule's collision partner undergo inelastic transitions.

The assumptions leading to this cross section for the scattering of two complex systems restrict its validity to relatively close encounters and to small angle scattering, hence intuitively one expects the predictions to be valid only for collisions of rather high energy. Indeed, the cross section is inversely proportional to the collision energy, and this behavior is known to be the limiting behavior at large collision energies for inelastic events occurring between particles with internal degrees of freedom.* Also, the structure of the collision partners, which is of a quantum mechanical nature, enters this theory only through the choice of the minimum excitation energy. This latter fact leads us to refer to cross sections based on the arguments above as classical results and is further delineated by the symbol Ru to indicate the use of Rutherford's formula for $I(\theta)$.

The somewhat more sophisticated application of classical mechanics to the manybody scattering problem introduced by Gryzinski[198] is labeled by G in the following discussion and of course is included in the category of classical results. The Gryzinski theory is not described here in detail. It is only mentioned to point out that it is quite similar to the Ru approach discussed above. The main point of departure is the possibility of including an appropriate velocity distribution for the "active" bound (electron) particle in the Gryzinski formulation.

In applying these classical models to dissociation events, it is usually not difficult to choose an appropriate minimum value for ΔE_α. Unfortunately all energy transfers greater than ΔE_α do not necessarily lead to dissociation. There is always the possibility of leaving the energy transferred to the molecule in excited electronic modes in a manner such that the molecule is not dissociated. The simple probability function described in a previous paragraph does not take this situation into account, and a method for correcting the classical model has yet to appear in the literature. Hence all available theoretical data based on these classical models contain this implicit approximation. The error may be large or small depending on the particular molecule being studied. Dissociation of H_2^+ has received by far the most theoretical attention and the error in this case is presumed to be small. Some reservation is probably justified even in the H_2^+ case, especially if one is interested in dissociation events occurring from H_2^+ initially in some excited vibrational state.

One questionable aspect of any classical theory is its use to predict the asymptotic behavior of the cross section at high collision energies. The failure of classical theories at high collision energies, for the case of electron excitation of a dipole-allowed transition, is dramatic in that an incorrect energy dependence is predicted.[198] The classical approach

* The cross section for an ion exciting a dipole-allowed transition in some target while the ion undergoes no inelastic transition proves the exception to this statement.

does predict the correct energy dependence, E^{-1}, at high collision energies for the dominant processes in the dissociation of a molecular system by collision with neutral particles, but one is still reluctant to accept the classical value for the coefficient of E^{-1}. The following qualitative argument adds to one's suspicions.[199] An estimate of the collision time at collision energies in the high-energy asymptotic range is such that the uncertainty in energy transfer, required by the uncertainty principle, can exceed or at least be a considerable fraction of ΔE_α. As indicated above, the classical formula for the coefficient of E^{-1} is inversely proportional to ΔE_α. The resulting importance of the choice of ΔE_α plus the required uncertainty in energy transfer for high-energy collisions is not a reassuring set of circumstances. Furthermore, an investigation of the high-energy asymptotic behavior of the first Born cross section for inelastic excitation events shows that the excitation energy plays no role at all in determining the leading coefficient.* The discussion of Section 7.5 returns to the problem of predicting the coefficient for E^{-1} by classical theories.

The first Born theory is now introduced. The total cross section for two heavy particles initially in the state i colliding to produce the final state f can be written as[201]

$$Q = (2\pi\hbar)^{-2} v_i^{-2} \int_0^{2\pi} d\varphi \int_{K_0}^{K_1} dK K \, |V_{if}|^2 \tag{7.6}$$

Here v_i is the magnitude of the relative collision velocity before scattering has taken place, and $\hbar K$ is the magnitude of the momentum transfer

$$K \equiv |\mathbf{K}| = |\mathbf{k}_i - \mathbf{k}_f| \tag{7.7}$$

The magnitude of the relative momentum of the collision partners before the collision is

$$k_i = |\mathbf{k}_i| = |\mu\hbar^{-1}\mathbf{v}_i| \tag{7.8}$$

and after the collision is

$$k_f = |\mathbf{k}_f| = |\mu\hbar^{-1}\mathbf{v}_f| = \sqrt{k_i^2 - 2\mu\hbar^{-2}\Delta E(i, f)} \tag{7.9}$$

where v_f is the magnitude of the relative collision velocity after the collision, μ is the reduced mass of the two scattering partners, and $\Delta E(i, f)$ is

---

* Reference 200 obtains an identity between the first Born and classical cross sections for vibrational dissociation. Hence this case may prove an exception to the rule just stated. The result obtained in ref. 200 depends, however, on a number of auxiliary approximations to the first Born theory. These approximations, such as a binary model of the collision, the assumed interaction potential, and the use of WKB-type wave functions, tend to enhance the analogy between classical and quantum theories.

the excitation energy. The angle φ is the azimuthal angle of the vector $\mathbf{K}$ in the space-fixed axis which is usually chosen with the Z axis colinear with $\mathbf{k}_i$. The magnitudes of the minimum and maximum momentum transfers consistent with energy conservation are

$$K_0 = k_i - k_f \tag{7.10}$$

and

$$K_1 = k_i + k_f \tag{7.11}$$

The matrix element appearing in Eq. 7.6 is

$$V_{if} = \langle \Psi_f \Phi_f \exp{(i\mathbf{k}_f \cdot \mathbf{r})} \mid V_{\text{int}} \mid \Psi_i \Phi_i \exp{(i\mathbf{k}_i \cdot \mathbf{r})} \rangle \tag{7.12}$$

where Ψ and Φ are the eigenfunctions describing the isolated collision partners, and the plane waves are the first Born approximations to the relative motion of the collision partners. At this point we assume Ψ to describe a diatomic molecule with nuclei having charges Z_α, Z_β, and Φ to describe an atom with nuclear charge Z_a. The operator V_{int} is usually taken as the static coulomb interaction between the particles (electrons and nuclei) making up one of the collision partners and the particles associated with the remaining collision partner. It is a simple matter to carry out the integration over the variable describing the relative position of the centers of mass of the two collision partners $\mathbf{r}$ and arrive at

$$V_{if} = 4\pi e^2 K^{-2} \langle \Psi_f \Phi_f \mid [\Sigma_n \exp{(i\mathbf{K} \cdot \mathbf{X}_n)} - Z_\alpha \exp{(i\mathbf{K} \cdot \mathbf{P}_\alpha)} \\ - Z_\beta \exp{(i\mathbf{K} \cdot \mathbf{P}_\beta)}] \times [\Sigma_m \exp{(-i\mathbf{K} \cdot \mathbf{Y}_m)} - Z_a] \mid \Psi_i \Phi_i \rangle \tag{7.13}$$

The vector $\mathbf{X}_n$ locates one of the electrons in the diatomic molecule with respect to the center of mass of the molecule's nuclei, $\mathbf{P}_\alpha$ locates the nucleus with charge Z_α in the same frame, $\mathbf{P}_\beta$ has the same significance for the nucleus with charge Z_β, and the sum over n includes all of the molecular electrons. The vector $\mathbf{Y}_m$ locates one of the electrons associated with the atom in a frame whose origin is at the center of mass of the atom, the atom has a nucleus with charge Z_a, and the m sum is over the atomic electrons. Introducing the Born matrix elements

$$\epsilon(\mathbf{K}) = \langle \Psi_f | \Sigma_n \exp{(i\mathbf{K} \cdot \mathbf{X}_n)} - Z_\alpha \exp{(i\mathbf{K} \cdot \mathbf{P}_\alpha)} - Z_\beta \exp{(i\mathbf{K} \cdot \mathbf{P}_\beta)} | \Psi_i \rangle \tag{7.14}$$

and

$$E(\mathbf{K}) = \langle \Phi_f | \Sigma_m \exp{(i\mathbf{K} \cdot \mathbf{Y}_m)} - Z_a | \Phi_i \rangle \tag{7.15}$$

the equality

$$V_{if} = 4\pi e^2 K^{-2} \epsilon(\mathbf{K}) E(-\mathbf{K}) \tag{7.16}$$

results. Substituting Eq. 7.16 into Eq. 7.6, and introducing atomic units, the cross section becomes

$$Q = 4v_i^{-2} \int_0^{2\pi} d\varphi \int_{K_0}^{K_1} dK \, K^{-3} |\epsilon(\mathbf{K})|^2 |E(-\mathbf{K})|^2 \tag{7.17}$$

Equation 7.17 also applies to the case of scattering by a structureless particle of charge Z_a if the identification $E(K) = -Z_a$ is made.

Results of the form of Eq. 7.17 are well known for the case of atom-atom scattering[201] and are simple enough to establish, in the framework outlined above, for the collision of two arbitrarily complex systems. An important aspect of Eq. 7.17 is the separation of the matrix element V_{if} into a product of matrix elements as shown by Eq. 7.16. The matrix element ϵ is identical to that encountered for the system in which the molecule undergoes the transition $i \to f$ on collision with an electron. The matrix element E has the same significance for the atom involved in the collision understudy. Hence the first Born theory for the scattering of complex systems has a close relationship to the problem of electron scattering by each of the collision partners. This situation is fortunate, since results from the study of the somewhat simpler problem of electron scattering can be used to understand the characteristics of ϵ and E occurring in Eq. 7.17. Also, the form of Eq. 7.17 motivates the inclusion of certain first Born theoretical data for electron-molecule scattering despite the exclusion of this topic from the present review. Much of the earlier work on electron-molecule collisions has been summarized.[202] References 203–209 are typical of the recent work in this area and refs. 210 and 211 present a critical study of several approximations that are often made in addition to the first Born approximation.

An interesting application of Eq. 7.17 has been proposed by Green.[212] He suggests that for anything but the simplest of systems, experimental studies of electron scattering provide much better data for $|\epsilon(\mathbf{K})|^2$ and $|E(-\mathbf{K})|^2$ than does theory. The experimental data, provided they exist, can then be used to construct the total cross section for the heavy particle system via Eq. 7.17. In this way one obtains a cross section independent of the approximations, usually associated with the lack of electronic eigenfunctions, encountered in treating complex scattering systems. Calculations of this type would be extremely useful in a careful assessment of the applicability of first Born theory to scattering between heavy particles. At the present time there appears to be a shortage of sufficiently precise experimental data, especially for large values of the magnitude of the momentum transfer K.[213] In selecting experimental data for this purpose care must be exercised to establish the "Born character" of the experimental data. The Born character is associated with the necessary condition that the experimental cross section differential in scattering angle be a function of only K for a given energy loss;[214] the sufficiency of this condition has not been established.

This idea has only been utilized to study two systems: proton excitation of the Schumann-Runge continuum of oxygen molecules[212] and the electronic excitation of H_2^+ on collision with helium.[213] The latter study uses a combination of theoretical and experimental data for the necessary Born matrix elements.

The Born-Oppenheimer separation of electron and nuclear motions[215] is one approximation that has been used universally in the treatment of scattering involving molecules. This approximation to Ψ_i and Ψ_f is now introduced and Eqs. 7.14 and 7.17 are rewritten to accommodate the associated notational conventions. For the case of a diatomic molecule in a discrete state, the molecular eigenfunction is approximated by

$$\Psi_i = \psi_i(. . .X_n. . .; R) F(\omega, R) \qquad (7.18)$$

Here ψ_i describes the electron motion in the space-fixed frame for a given value of the internuclear vector $R = P_\alpha - P_\beta$, and i describes the initial electronic eigenstate. The relative motion of the nuclei is characterized by the eigenfunction $F(\omega, R)$, and the initial eigenstate is labeled ω. Since Ψ_i is a bound state, both ψ_i and $F(\omega, R)$ are normalized to unity. The total wave function for the final state is written in a similar fashion:

$$\Psi_f = \psi_f(. . .X_n. . .; R) F(\kappa, R) \qquad (7.19)$$

The eigenfunction ψ_f may either represent a discrete or continuum state of motion for the electrons, but it is assumed to be a discrete state in the following development to simplify the notation. Since we are dealing with dissociative collisions, $F(\kappa, R)$ necessarily represents a state of nuclear motion with momentum $\hbar\kappa$ in the continuum. The boundary condition for R having a large magnitude requires $F(\kappa, R)$ to become a plane wave plus an incoming spherical wave. The normalization

$$\int dR\, F^*(\kappa', R)\, F(\kappa, R) = \delta(\kappa - \kappa') \qquad (7.20)$$

is chosen. The energy of dissociation is $\hbar^2(2M)^{-1}|\kappa|^2$, where M is the reduced mass of the dissociating fragments. A number of examples of the explicit form for $F(\kappa, R)$ and its application in first Born studies of dissociative collisions[216,217] can be found in the literature.

The use of Eqs. 7.18 and 7.19 allows the matrix element defined by Eq. 7.14 to be written as

$$\epsilon(K, \kappa) = \int dR\, F^*(\kappa, R)\, \epsilon_{el}(K, R)\, F(\omega, R) \qquad (7.21)$$

where the electronic Born matrix element is

$$\epsilon_{el}(K, R) = \int d\tau\, \psi_f^*(. . .X_n. . .; R)[\Sigma_n \exp(iK \cdot X_n) - Z_\alpha \exp(iK \cdot P_\alpha) - Z_\beta \exp(iK \cdot P_\beta)]\psi_i(. . .X_n. . .; R) \qquad (7.22)$$

The quantity κ has been added to the argument of ϵ to emphasize the dependence on this quantity. The internuclear separation R is added to the argument of ϵ_{el} for the same reason. Because of the nature of the normalization of $F(\kappa, \mathbf{R})$, the cross section, Eq. 7.17, is now differential with respect to $d\kappa$, hence we must rewrite Eq. 7.17 as

$$\frac{dQ}{d\kappa} = 4v_i^{-2} \int_0^{2\pi} d\varphi \int_{K_0}^{K_1} dK \, K^{-3} |\epsilon(\mathbf{K}, \kappa)|^2 |E(-\mathbf{K})|^2 \qquad (7.23)$$

where Eq. 7.21 has replaced Eq. 7.14. Equation 7.23 is differential both in the energy of dissociation and the direction of dissociation associated with the element of solid angle $d\Omega(\kappa)$ for the vector κ. Another differential cross section of interest is

$$\frac{dQ}{d\Omega(\kappa)} = 4v_i^{-2} \int_0^{\kappa^{\ddagger}} \kappa^2 \, d\kappa \int_0^{2\pi} d\varphi \int_0^{K_1} dK \, K^{-3} |\epsilon(\mathbf{K}, \kappa)|^2 |E(-\mathbf{K})|^2 \qquad (7.24)$$

where $\kappa^{\ddagger}$ is the maximum magnitude of dissociation momentum allowed by energy conservation requirements. Equation 7.24 is differential in the angle of dissociation of the molecule but not in the dissociation momentum (energy) or the ordinary scattering angle θ utilized in Eq. 7.5. There is the possibility of another differential cross section $dQ/\kappa^2 \, d\kappa$ which resolves the dissociation momentum but not the direction of the dissociation associated with $d\Omega(\kappa)$. This type of differential cross section has not been often studied theoretically for heavy particle collisions, although it is of interest in electron-molecule scattering. The cross section differential in the scattering angle θ is as yet of little interest in dissociative collisions.

One often requires a cross section that has been averaged over all degenerate levels of the initial state(s) and summed over all degenerate levels of the final state(s). Most degeneracies will, of course, be removed if enough detail is incorporated into the theory. Effects associated with spin-orbit interactions and other relativistic phenomena are examples of details often ignored. We presume these unenumerated "small effects" are of little importance for the processes under consideration, but it should not be inferred that this is true for all situations. The summing-averaging process for the electronic degrees of freedom is well documented in the electron-atom literature. Treatment of the rotational-vibrational degrees of freedom has been discussed in a number of places.[202,207]

The summing-averaging process commutes with the dK integration when the model described in the preceding paragraph is adopted. Writing the summed-averaged total cross section as $\bar{Q}$, the result is

$$\begin{aligned}
\bar{Q} &= 4v_i^{-2} \int_0^{\kappa^{\ddagger}} \kappa^2 \, d\kappa \int_0^{2\pi} d\varphi \int_{K_0}^{K_1} dK \, K^{-3} \overline{|\epsilon(\mathbf{K}, \kappa)|^2} \; \overline{|E(-\mathbf{K})|^2} \\
&= 8\pi v_i^{-2} \int_0^{\kappa^{\ddagger}} \kappa^2 \, d\kappa \int_{K_0}^{K_1} dK \, K^{-3} \overline{|\epsilon(\mathbf{K}, \kappa)|^2} \; \overline{|E(-\mathbf{K})|^2} \qquad (7.25)
\end{aligned}$$

The last step follows because the summed and averaged Born matrix elements $\overline{|\epsilon(\mathbf{K}, \kappa)|^2}$ and $\overline{|E(-\mathbf{K})|^2}$ depend only on $|\mathbf{K}|$ and $|\kappa|$. See ref. 211 for a proof of the above statement for $\overline{|\epsilon(\mathbf{K}, \kappa)|^2}$. The effects of rotational-vibrational coupling are included in ref. 211 and can be compared with results where this type of coupling is ignored; see refs. 206 and 209 for an example of the latter treatment.

The approximations commonly used to evaluate Eqs. 7.23–7.25 are enumerated in the following paragraphs. They are identified with a number that is used in Section 7.4 to describe the contents of the theoretical papers under review. An attempt is made to assess qualitatively the error associated with the various approximations.

The discussions of the approximations are truncated in two respects. First, the approximations are stated in a rather schematic way and do not display details, such as equations, resulting from the various approximations. Also, the evaluation of errors is necessarily not quantitative due to the differences between scattering systems. The only effective way to add to this outline is to refer the reader to the original literature. The outline of approximations is not intended to provide this guide to the literature; the guide is provided by the use of Table 7.3 in conjunction with the following outline. If the references cited in discussing a particular approximation prove inadequate, a survey of Table 7.3 provides a list of papers that have used the approximation in question. The resulting list is not long. The utility of a particular paper cannot be guaranteed, but a paper discussing the application of more than one approximation to the same scattering system, if available, is usually the best place to start.

Approximation 1. The introduction of the approximations implied by Eqs. 7.18 and 7.19 has been made in all scattering work under consideration in this review. Although no critical test of the Born-Oppenheimer approximation in the framework of this scattering problem has been performed, there seems to be little suspicion that this is a source of difficulty. The necessity of including nonadiabatic effects can be anticipated as experimental techniques become more refined.

However, an important observation concerning the use of Eq. 7.19 has been made.[218,219] In discussing certain differential cross sections associated with Eqs. 7.23 and 7.24 it was shown that Eq. 7.19 did not satisfy appropriate boundary conditions when two or more molecular states were degenerate at infinite internuclear separations. This always occurs for a homopolar diatomic molecule. It proved necessary to replace Eq. 7.19 with a superposition of certain molecular states with this type of degeneracy. The scattering amplitudes from these various states were then found to interfere in forming the differential cross section defined by Eq. 7.23. However, these interference terms integrate to zero in constructing the total cross section defined by Eq. 7.25. No detailed theoretical study of this effect has yet been performed.

Approximation 2. The precise evaluation of Eq. 7.22 is one of the most difficult parts of the computational problem. It is the same as the problem encountered in electron-atom studies except for the additional independent variable $\mathbf{R}$. Some studies of the evaluation of Eq. 7.22 using various approximations to ψ_i and ψ_f have been carried out, and sizable errors, tens of percents in the total cross section, have been encountered[220] for the electron-molecule case where small momentum transfers are emphasized. It can be argued that approximate wave functions, such as those constructed from a variational approach, would lead to smaller errors for processes in which larger momentum transfers dominate the total cross section. It seems probable that, with reasonable care for simple systems, approximate evaluation of Eqs. 7.15 and 7.22 leads to results of sufficient accuracy for the ordinary scattering situation. For more complicated systems, the proposal made by Green[212] discussed above offers a promising alternative for evaluating Eqs. 7.23–7.25 if reliable first Born cross sections for heavy particle collisions are desired.

Approximation 3. One of the most common approximations made in early electron-molecule scattering studies[202] is that ϵ_{el} is independent of R:

$$\epsilon_{el}(\mathbf{K}, \mathbf{R}) \cong \epsilon_{el}(\mathbf{K}, \mathbf{R}^*) \qquad (7.26)$$

where $\mathbf{R}^*$ has a magnitude characteristic of the problem. Also, the rotational and vibrational motions in the final state are assumed to be uncoupled. The cross section is then summed over all rotational modes associated with the final vibrational state and averaged over the degenerate rotational levels of the initial vibrational state. In addition, the rotational sum is assumed to commute with the dK integration. As a consequence of these approximations and the completeness of the uncoupled rotational eigenfunctions it is possible to use the closure property of these rotational functions to simplify the Born matrix element. The resulting cross section consists of the well-known Franck-Condon factor multiplied by an unweighted average over all orientations of the cross section for the fixed value of $\mathbf{R}^*$. The Franck-Condon factor is usually ignored or summed using the closure property of the uncoupled final state vibrational eigenfunctions.[202] This approximation may be reliable in a given case, but should not be used uncritically. It is, of course, most accurate when the molecule is known to be initially in a rotational-vibrational state that does not allow the magnitude of $\mathbf{R}$ to vary significantly.

Approximation 4. Another approximation also utilizes the closure property of the complete set of rotational-vibrational eigenfunctions associated with the final electronic state:

$$\Sigma_{\omega'} F(\omega', \mathbf{R}) F^*(\omega', \mathbf{R}') + \int d\kappa\, F(\kappa, \mathbf{R}) F^*(\kappa, \mathbf{R}') = \delta(\mathbf{R} - \mathbf{R}') \quad (7.27)$$

where ω' refers to a discrete state in the final electronic state and κ is defined in connection with Eqs. 7.19 and 7.20. Equation 7.27 is only used in conjunction with Eq. 7.25, and its use involves two distinct approximations. The first approximation is to extend the integration over $d\kappa$ from $\kappa^{\ddagger}$ to infinity and add to the $d\kappa$ integral the sum over ω' as it appears in Eq. 7.27. This has the effect of converting the $d\kappa$ integral in Eq. 7.25 into the so-called closure sum; the closure sum is defined by the sum plus integral appearing in Eq. 7.27. The second approximation is the interchange of the closure sum with the dK integration. A considerable simplification of the $|\epsilon(\mathbf{K}, \kappa)|^2$ matrix element is now possible.

The first approximation requires us to interpret $\bar{Q}$ as the cross section which includes all states of nuclear motion in the final electronic state. To apply this result to dissociative collisions it is necessary to subtract terms from $\bar{Q}$ that correspond to excitation of discrete modes of nuclear motion in the final electronic state.

The second approximation reduces the meaning of $\Delta E(i, f)$ to a parameter in the theory. Some investigation of the effect of the choice of $\Delta E(i, f)$ on the cross section, which includes all states of nuclear motion in the final electronic state, has appeared in the literature. Two different choices of the parameter have been used which give results that differ by only a small amount at relatively large collision energies.[217] A critical study[210] of this use of Eq. 7.27 for the case of electron-molecule scattering has shown it to be asymptotically correct at high collision energies. It is obvious from ref. 210 that the use of this approximation to evaluate Eq. 7.25 also gives the correct first term of a high collision energy asymptotic expansion of $\bar{Q}$. (The reason for this relationship between the exact first Born cross section and the results based on the approximation being discussed is associated with the fact that the first term in a high-energy asymptotic expansion of the $B1$ cross section is independent of energy conservation requirements.) Hence the error associated with this approximation is known to vanish asymptotically, although general knowledge of the error at finite collision energies is not available. However, it seems clear that this approximation, unless misused in the extreme, is always as good and usually better than Approximation 3 since the rather restrictive approximation of Eq. 7.26 is avoided.

Approximation 5. Another approximation that is often used along with Approximation 4 is to ignore the subtraction of the terms corresponding to ω' from the closure sum over $\bar{Q}$. As discussed in Approximation 4, these terms correspond to exciting discrete states of nuclear motion in the final electronic state. Ignoring these terms may or may not introduce serious inaccuracies.

The following argument[211] is usually sufficient to estimate the situation. If the initial rotational-vibrational eigenfunction is essentially zero over the range of R for which the final electronic state shows attractive, that is, bonding, character, the omission of the subtraction terms should introduce little error. Otherwise, rather significant errors can be expected. Obviously, omitting the subtraction terms produces an upper bound, in the limit of large collision energies, to the Born cross section for dissociation.

Approximation 6. The discussion of Approximations 4 and 5 presumed that the dissociative event is due to electronic excitation of the molecule. Approximation 4 can also be used when electronic excitation has not taken place. This would produce a $\bar{Q}$ that includes all possible rotational-vibrational excitation and dissociative events, provided the elastic channel is correctly subtracted from the sum. If one makes an approximation analogous to Approximation 5 for this type of event, it seems clear the error is likely to be quite large unless the discrete sum in Eq. 7.27 consists of a very few states.

Approximation 7. Closure over the electronic degrees of freedom of the molecule has also proven to be of use. This gives a $\bar{Q}$ which is a composite of all electronic states minus the ones that have been subtracted from the closure sum. When used in conjunction with the electronic states of the molecule it is necessary to subtract the elastic electronic channel and sometimes, also, excited events. As indicated in Table 7.1, the subtraction of the elastic channel is indicated by a prime on the sum, and so forth. In one case,[205] the closure property that was used corresponded to the completeness of functions associated with a symmetry property of the associated degree of freedom. Specifically, a sum over all electronic orbitals with either sigma or pi symmetry has been used. This special type of sum is indicated in the tables by appending the appropriate subscript to the summation sign. When used in conjunction with Approximation 4, this closure result is ordinarily only used to estimate cross sections that contribute a small amount to other cross sections obtained by a more accurate technique. Provided the closure sum over electronic degrees of freedom does not include unwanted excitation events, the approximate cross section with the usual choice of "excitation energy" $\Delta E(i, f)$ can be expected to be too large, although the error vanishes in the limit of high collision energies by the arguments outlined in Approximation 4. If undesired excitations are included in the closure argument, the resulting estimate is too large even in the high-energy limit.

Approximation 8. If the molecule is dissociated by a particle with internal degrees of freedom, it is necessary to consider simultaneous excitation of the two systems. Often, when considering a given dissociative excitation, one wishes to construct a cross section that includes all possible events in the molecule's collision partner. This situation can be conveniently treated using closure arguments. It is customary to treat separately the event in which the molecule's collision partner is left unexcited. As in the other closure approximations, this approach is probably accurate at very high collision energies, in the v_i^{-2} asymptotic range, but produces serious errors at smaller collision energies. Correction terms are easily derived and must be used at the lower energies. These correction terms indicate that the error is positive but depends on the assumed excitation energy, which is a parameter in all of these closure approximations as discussed in Approximation 4. When the closure approximation discussed in this paragraph is used with Approximation 7 the results are limited to the v_i^{-2} asymptotic energy range.

Approximation 9. This approximation defines the use of Approximation 8 when the correction terms discussed in the preceding paragraph are omitted.

Approximation 10. An approximation that has proved useful in evaluating Eqs. 7.23 and 7.24 is the so-called reflection method.[221-223] This model can be constructed from the analysis given above by replacing $F(\kappa, \mathbf{R})$ with a three-dimensional Dirac delta function.* The argument of the delta function is $\mathbf{R} - \mathbf{R}^*$. The vector $\mathbf{R}^*$ is collinear to the vector κ. The magnitude of $\mathbf{R}^*$ is chosen as the intersection of the line representing the energy associated with $F(\kappa, \mathbf{R})$, on a plot of energy verus R, with the curve representing the appropriate adiabatic electronic eigenenergy as a function of R. (The reflection method probably should not be used if this definition of $\mathbf{R}^*$ is not single valued.[224]) The normalization of the delta function is chosen to insure the satisfaction of Eq. 7.27 *provided the sum over ω' is absent.* The reflection model gives a particularly simple

---

* Usually the reflection model is equated to the use of a Dirac delta function in the magnitude of $\mathbf{R}$ and the remaining variables are not discussed. If the functional form $\delta(R - R^*)$ is assumed for the radial part of $F(\kappa, \mathbf{R})$, one is led to the description of the reflection model given here. Investigation of the form of $F(\kappa, \mathbf{R})$, plus recognition of the fact that the choice of a radial function independent of rotational angular momentum, such as $\delta(R - R^*)$, is not consistent with defining a phase shift dependent on rotational angular momentum, proves the assertion.

result. The information required to evaluate $\bar{Q}$ of Eq. 7.25 in the closure approximation defined as Approximation 4 is sufficient to evaluate the reflection approximation to the cross sections differential in $d\kappa$, $\kappa^2 d\kappa$, or $d\Omega(\kappa)$.

The interpretation of cross sections differential in $\kappa^2 \, d\kappa$, as given by the reflection model, in terms of the "reflection" of the radial part of $F(\omega, \mathbf{R})$ (see Eq. 7.18) on the final electronic state potential curve is well documented elsewhere.[221–223] Discussion in the papers cited by refs. 221–223 leads to the conclusion that the distribution in κ, equivalently the dissociation energy, predicted by the reflection model, is narrower than the result based on the correct $F(\kappa, \mathbf{R})$. This phenomenon has been observed in photo dissociation,[225–226] and the well-known analogy between photon scattering and high-energy electron scattering (i.e., any structureless charged particle) has been used to discuss this point in dissociation by electron scattering.[224] The examples studied in photon-molecule scattering indicated that the errors associated with the reflection model are not large.

The reflection model as applied to this three-dimensional analysis asserts that the direction of dissociation, defined by the direction of κ, is collinear with the direction of the internuclear axis $\mathbf{R}$. The general form of $F(\kappa, \mathbf{R})$, of course, does not require κ to be collinear with $\mathbf{R}$. The term "axial recoil" is often used to describe this aspect of the reflection model;[227] it should be clear that the axial recoil approximation does not imply the reflection approximation. We are not aware of any detailed tests for the axial recoil approximation, although the conjecture[227] that the angular distribution $d\Omega(\kappa)$ based on $F(\kappa, \mathbf{R})$ is more isotropic than the one predicted by the axial recoil aspect of the reflection model seems reasonable.

Cross sections differential in $d\kappa$, $\kappa^2 \, d\kappa$, or $d\Omega(\kappa)$ occurring in heavy particle scattering have only been calculated in the reflection model. Hence little that is appropriate to the present problem can be said about errors associated with this approximation other than the implications discussed in the preceding two paragraphs, plus the addition of the following comment. A study of total dissociation cross sections based on the reflection model, as defined in ref. 224, has shown this approximation to be grossly in error for an example of electron-molecule scattering when excitation of both discrete and continuum modes of nuclear motion are possible.[211] This is in part due to the normalization discussed above of the Dirac delta function. Large errors in the differential cross sections can then be inferred if the final electronic state potential curve indicates some possibility of exciting discrete states of nuclear motion. The discussion in Approximation 5 is appropriate for deciding whether or not the error will be large. In the event that there are no discrete states of nuclear motion in the final electronic state, an equivalence of the total dissociation cross section calculated via Approximation 4 and by the reflection approximation has been established.[217] In this case the reflection model probably gives reasonably accurate differential cross sections.

7.4 LITERATURE SURVEY

Table 7.2 contains a survey of the particular dissociative collision events for which theoretical data are available and includes references to the authors who carried out the investigations. Information concerning the collision energy range studied for the various processes is not provided and on this point reference to the original paper is required. Generally speaking, most investigations cover an energy range bounded on the low-energy side by threshold, or an energy below which the theory is probably not valid, and extend to an energy range where the cross section has its high-energy asymptotic behavior. Some studies, however, cover a more restricted range of collision energies. Classical cross sections, for example, have been calculated only for the high-energy asymptotic range. It should also be pointed out that many data have been listed for molecular collisions with protons. The indicated paper may in fact discuss the case of molecular scattering of an arbitrary structureless charged particle, or the paper may be concerned with electron scattering. It is appropriate to include the latter case in the present review provided that the first Born approximation is used and the dissociation mechanism is available to proton scattering. It is well known that the cross section for electron scattering at all collision energies is sufficient to generate the analogous cross section for a structureless particle with arbitrary mass and charge if these conditions are met.[201]

Table 7.3 is an amplification of the contents of Table 7.2. The cross sections treated in each paper are listed, as are the theory and approximations pertinent to the various studies. Those papers giving comparison with experiment and/or with independent theoretical results are indicated, and amplifying comments are made where necessary. The listing of approximations for the various papers cannot necessarily be interpreted as a value judgment of the paper. The degree of error associated with a given approximation is a function of the system under consideration and the intent of the author's work mitigates the importance of the approximation. Reference to the original paper is essential for an accurate perspective.

TABLE 7.2

This table contains a listing of the various dissociative collisions of interest to this review which have received theoretical attention. The ordering is such that results for the simplest molecule appear first and these entries are followed by results for progressively more complicated molecules. Within a group of entries for a given molecule, the ordering is in terms of the simplicity of the molecule's collision partner. Column 1 defines the molecule and its state prior to the collision while column 2 provides the same information for the molecule's collision partner. Column 3 defines the dissociative modes of the molecule and column 4 gives the final states of the molecule's collision partner. The symbols used to characterize the states of the collision partners are defined in Table 7.1. The pertinent reference numbers are given in column 5.

1	2	3	4	5
$H_2^+(0;2.0)$	H^+	$H_2^+(D;2.0)$	H^+	230,229
$H_2^+(0;2.0)$	H^+	$H_2^+(I;2.0)$	H^+	230,229
$H_2^+(0;2.0)$	H^+	$2H^+$	$H(1s)$	230
$H_2^+(0;2.0)$	H^+	$H_2^+(2p\sigma_u;2.0)$	H^+	244,204,205,227
$H_2^+(0;2.0)$	H^+	$H_2^+(2p\pi_u;2.0)$	H^+	244,204,205
$H_2^+(0;2.0)$	H^+	$H_2^+(2s\sigma_g;2.0)$	H^+	204,205
$H_2^+(0;2.0)$	H^+	$H_2^+(3d\pi_g;2.0)$	H^+	205
$H_2^+(0;1.4)$	H^+	$H_2^+(2p\sigma_u;1.4)$	H^+	204
$H_2^+(0;3.2)$	H^+	$H_2^+(2p\sigma_u;3.2)$	H^+	204
$H_2^+(0;1-5)$	H^+	$H_2^+(2p\sigma_u;1-5)$	H^+	205
$H_2^+(0;1-5)$	H^+	$H_2^+(2p\pi_u;1-5)$	H^+	205
$H_2^+(0;1-5)$	H^+	$H_2^+(\Sigma''_\sigma;1-5)$	H^+	205
$H_2^+(0;1-5)$	H^+	$H_2^+(\Sigma'_\pi;1-5)$	H^+	205

TABLE 7.2

1	2	3	4	5
$H_2^+(0;\nu J)$	H^+	$H_2^+(2p\sigma_u;\Sigma)$	H^+	205,217,231,211,224
$H_2^+(0;\nu J)$	H^+	$H_2^+(2p\pi_u;\Sigma)$	H^+	231
$H_2^+(0;\nu J)$	H^+	$H_2^+(\Sigma';\Sigma)$	H^+	205
$H_2^+(0;\nu J)$	H^+	$H_2^+(\Sigma'';\Sigma)$	H^+	231
$H_2^+(0;\nu J)$	H^+	$H_2^+(V;\Sigma)$	H^+	205
$H_2^+(0;2.0)$	$H(0)$	$H_2^+(D;2.0)$	$H(\Sigma)$	196
$H_2^+(0;2.0)$	$H(0)$	$H_2^+(I;2.0)$	$H(\Sigma)$	196
$H_2^+(0;2.0)$	$H(0)$	$H_2^+(V;2.0)$	$H(\Sigma)$	196
$H_2^+(0;\nu J)$	$H(0)$	$H_2^+(D;\Sigma)$	$H(0)$	37
$H_2^+(0;\nu J)$	$H(0)$	$H_2^+(D;\Sigma)$	$H(\Sigma')$	37
$H_2^+(0;\nu J)$	$H(0)$	$H_2^+(I;\Sigma)$	$H(0)$	37
$H_2^+(0;\nu J)$	$H(0)$	$H_2^+(I;\Sigma)$	$H(\Sigma')$	37
$H_2^+(0;\nu J)$	$H(0)$	$H_2^+(V;\Sigma)$	$H(\Sigma)$	37
$H_2^+(0;2.0)$	$H(0)$	$H_2^+(2p\sigma_u;2.0)$	$H(\Sigma)$	217
$H_2^+(0;1.4-10)$	$H(0)$	$H_2^+(2p\sigma_u;1.4-10)$	$H(0)$	63
$H_2^+(0;R)$	$H(0)$	$H_2^+(2p\sigma_u;\kappa)$	$H(0)$	63,111
$H_2^+(0;\bar{R})$	$H(0)$	$H_2^+(2p\sigma_u;\kappa)$	$H(0)$	63,40,111
$H_2^+(0;\bar{R})$	$H(0)$	$H_2^+(2p\pi_u;\kappa)$	$H(0)$	40
$H_2^+(0;R)$	$H(0)$	$H_2^+(2p\sigma_u;\kappa)$	$H(\Sigma)$	40

TABLE 7.2

1	2	3	4	5
$H_2^+(0;\overline{R})$	H(0)	$H_2^+(2p\sigma_u;\kappa)$	H(Σ)	40
$H_2^+(0;\nu J)$	H(0)	$H_2^+(2p\sigma_u;\Sigma)$	H(Σ)	217,235
$H_2^+(0;\nu J)$	H(0)	$H_2^+(2p\sigma_u;\Sigma)$	H(0)	217,235
$H_2^+(0;\nu J)$	$H_2(0;1.4)$	$H_2^+(2p\sigma_u;\Sigma)$	$H_2(\Sigma;1.4)$	234
$H_2^+(0;\nu J)$	$H_2(0;1.4)$	$H_2^+(2p\pi_u;\Sigma)$	$H_2(\Sigma;1.4)$	234
$H_2^+(0;\nu J)$	$H_2(0;1.4)$	$H_2^+(\Sigma';\Sigma)$	$H_2(\Sigma;1.4)$	234
$H_2^+(0;\nu J)$	He(0)	$H_2^+(D;\Sigma)$	He(0)	37
$H_2^+(0;\nu J)$	He(0)	$H_2^+(D;\Sigma)$	He(Σ')	37
$H_2^+(0;\nu J)$	He(0)	$H_2^+(I;\Sigma)$	He(0)	37
$H_2^+(0;\nu J)$	He(0)	$H_2^+(I;\Sigma)$	He(Σ')	37
$H_2^+(0;\nu J)$	He(0)	$H_2^+(V;\Sigma)$	He(Σ)	37
$H_2^+(0;\nu J)$	He(0)	$H_2^+(2p\sigma_u;\Sigma)$	He(0)	236
$H_2^+(0;\nu J)$	He(0)	$H_2^+(V;\Sigma)$	He(0)	236
$H_2^+(0;\nu J)$	He(0)	$H_2^+(2p\sigma_u+2s\sigma_g+2p\pi_u;\Sigma)$	He(0)	213
$H_2^+(0;\nu J)$	He(0)	$H_2^+(2p\sigma_u+2s\sigma_g+2p\pi_u;\Sigma)$	He(I)	213
$H_2^+(0;\nu J)$	He(0)	$H_2^+(2p\sigma_u+2s\sigma_g+2p\pi_u;\Sigma)$	He(2^1S$+2^1$P$+3^1$P+I)	213
$H_2^+(0;\nu J)$	He(0)	$H_2^+(2p\sigma_u+2s\sigma_g+2p\pi_u;\Sigma)$	He(Σ')	213
$H_2^+(0;\nu J)$	He(0)	$H_2^+(\Sigma';\Sigma)$	He(Σ)	213
$H_2^+(0;R)$	He(0)	$H_2^+(2p\sigma_u;\kappa)$	He(2^1P)	219

TABLE 7.2

1	2	3	4	5
$H_2^+(0;\nu J)$	He(0)	$H_2^+(2p\sigma_u;\kappa)$	He(0)	240
$H_2^+(0;\overline{R})$	He(0)	$H_2^+(V;\kappa)$	He(0)	240
$H_2^+(0;\overline{R})$	He(0)	$H_2^+(V;\kappa)$	He(Σ)	240
$H_2^+(0;2.0)$	N(0)	$H_2^+(D;2.0)$	N(Σ)	196
$H_2^+(0;2.0)$	N(0)	$H_2^+(I;2.0)$	N(Σ)	196
$H_2^+(0;2.0)$	N(0)	$H_2^+(V;2.0)$	N(Σ)	196
$H_2^+(0;\nu J)$	N(0)	$H_2^+(D;\Sigma)$	N(0)	37
$H_2^+(0;\nu J)$	N(0)	$H_2^+(D;\Sigma)$	N(Σ')	37
$H_2^+(0;\nu J)$	N(0)	$H_2^+(I;\Sigma)$	N(0)	37
$H_2^+(0;\nu J)$	N(0)	$H_2^+(I;\Sigma)$	N(Σ')	37
$H_2^+(0;\nu J)$	N(0)	$H_2^+(V;\Sigma)$	N(Σ)	37
$H_2^+(0;2.0)$	O(0)	$H_2^+(D;2.0)$	O(Σ)	196
$H_2^+(0;2.0)$	O(0)	$H_2^+(I;2.0)$	O(Σ)	196
$H_2^+(0;2.0)$	O(0)	$H_2^+(V;2.0)$	O(Σ)	196
$H_2^+(0;2.0)$	Ar(0)	$H_2^+(D;2.0)$	Ar(Σ)	196
$H_2^+(0;2.0)$	Ar(0)	$H_2^+(I;2.0)$	Ar(Σ)	196
$H_2^+(0;2.0)$	Ar(0)	$H_2^+(V;2.0)$	Ar(Σ)	196
$H_2^+(0;\nu J)$	Ar(0)	$H_2^+(D;\Sigma)$	Ar(0)	37
$H_2^+(0;\nu J)$	Ar(0)	$H_2^+(D;\Sigma)$	Ar(Σ')	37

TABLE 7.2

1	2	3	4	5
$H_2^+(0;\nu J)$	Ar(0)	$H_2^+(I;\Sigma)$	Ar(0)	37
$H_2^+(0;\nu J)$	Ar(0)	$H_2^+(I;\Sigma)$	Ar(Σ')	37
$H_2^+(0;\nu J)$	Ar(0)	$H_2^+(V;\Sigma)$	Ar(Σ)	37
$H_2^+(0;R)$	Ar(0)	$H_2^+(2p\sigma_u,2p\pi_u,2s\sigma_g;R)$	Ar(0)	219
$H_2^+(0;R)$	Ar(0)	$H_2^+(2p\sigma_u,2p\pi_u,2s\sigma_g;\kappa)$	Ar(0)	219
$H_2(0;\nu J)$	H^+	$H_2(V;\Sigma)$	H^+	245,200,246
$H_2(0;1.4)$	H^+	$2H^+$	$H^-(0)$	241
$H_2(0;\nu J)$	H(0)	$H_2(V;\Sigma)$	H(0)	245,246
$O_2(0;2.3)$	H^+	$O_2(^3\Sigma_u^-;2.3)$	H^+	212,247,248
$O_2(0;\nu J)$	H^+	$O_2(^3\Sigma_u^-;\Sigma)$	H^+	212
$NaCl(0;\nu J)$	H^+	$NaCl(V;\Sigma)$	H^+	200
$HgH(0;\nu J)$	H^+	$HgH(V;\Sigma)$	H^+	200

TABLE 7.3

This table contains a list of the various papers roughly in chronological order. The first column contains the reference number for the particular paper. The second column indicates the particular framework used to treat the collision listed under Process and the third column lists the additional approximations that were used. Amplifying remarks are given in the last column. Reference to Section 7.3 will give information concerning columns 2 and 3 and reference to Table 7.1 amplifies the notation, especially concerning column 4. The letter c appears in parenthesis after the reference number for those papers giving comparisons between their results and other theoretical and/or experimental results.

Reference	Theory	Approximations	Process	Remarks
196	C	R	(a) $H_2^+(0;2.0) + X(0) \rightarrow$ $H_2^+(D;2.0) + X(\Sigma)$	$X = H, N, O, Ar$ (a) Incorporates screening into the classical forma-tion.
	C	R	(a) $H_2^+(0;2.0) + X(0) \rightarrow$ $H_2^+(I;2.0) + X(\Sigma)$	
	C	R	(a) $H_2^+(0;2.0) + X(0) \rightarrow$ $H_2^+(V;2.0) + X(\Sigma)$	
245,246	B1	1,2,‡	(a) $H_2(0;\nu J) + H^+ \rightarrow$ $H_2(V;\Sigma) + H^+$	(a) The first paper treats $\nu=0$, J=0 and the second paper estimates the dissociation cross section for high ν and J=0
	B1	1,2,‡	(a) $H_2(0;\nu J) + H(0) \rightarrow$ $H_2(V;\Sigma) + H(0)$	‡ Approximates the interaction between H_2 and H, H^+. Also uses several approximations to the rotational-vibrational wave functions.
230	B1	*	$H_2^+(0;2.0) + H^+ \rightarrow$ $H_2^+(D;2.0) + H^+$	* Scales B1 cross sections for analogous processes in He^+-H^+ and H-H^+ scattering.

TABLE 7.3

Reference	Theory	Approximation	Process	Remarks
	B1	*	$H_2^+(0;2.0) + H^+ \rightarrow$ $H_2^+(1;2.0) + H^+$	
	B1	‡	$H_2^+(0;2.0) + H^+ \rightarrow$ $2H^+ + H(1s)$	‡ Scales Schiff results for H+H$^+$ charge transfer.
244(c)	B1	1,3	$H_2^+(0;2.0) + H^+ \rightarrow$ $H_2^+(2p\sigma_u;2.0) + H^+$	Note that approximation 2 was not used.
	B1	1,3	$H_2^+(0;2.0) + H^+ \rightarrow$ $H_2^+(2p\pi_u;2.0) + H^+$	
229(c)	C	G	$H_2^+(0;2.0) + H^+ \rightarrow$ $H_2^+(D;2.0) + H^+$	Uses the linear combination of atomic orbits H_2^+ wave function to calculate the velocity distribution required by the Gryzinski theory. Also gives electron scattering data.
	C	G	$H_2^+(0;2.0) + H^+ \rightarrow$ $H_2^+(1;2.0) + H^+$	
247,248	‡		$O_2(0;2.282) + H^+ \rightarrow$ $O_2(^3\Sigma_u^-;2.282) + H^+$	‡ Based on a two-state impact parameter formation.
204(c)	B1	1,3	(a) $H_2^+(0;1.4,2.0,3.2) + H^+ \rightarrow$ $H_2^+(2p\sigma_u;1.4,2.0,3.2) + H^+$	(a) Tabulates $\overline{\lvert \epsilon(\mathbf{K},\kappa)\rvert^2}/\kappa^2$ in the indicated approximations. A few examples of $\overline{Q}$, using approximations 1,4,5, for e$^-$ scattering are given.
	B1	1,3	(b) $H_2^+(0;2.0) + H^+ \rightarrow$ $H_2^+(2p\pi_u;2s\sigma_g,2.0) + H^+$	(b) Tables of $\overline{\lvert \epsilon(\mathbf{K},\kappa)\rvert^2}/K^2$ and graphs of $\overline{Q}$, using approximations 1,3, for e$^-$ scattering are given.

TABLE 7.3

Reference	Theory	Approximation	Process	Remarks
227	B1	1,2,3,10	(a) $H_2^+(0;2.0) + H^+ \to$ $H_2^+(2p\sigma_u;\kappa) + H^+$	(a) The cross section $dQ/d\kappa$ is given for high energy e^- scattering.
224	B1	1,‡	(a) $H_2^+(0;\nu J) + H^+ \to$ $H_2^+(2p\sigma_u;\Sigma) + H^+$	(a) $\nu=0(1)18; J=0$ ‡ Approximations were made to both the H_2^+ potential curves and vibrational wave functions, also one term in his representation of the total cross section was approximated.
240	C	‡	(a) $H_2^+(0;\nu J) + He(0) \to$ $H_2^+(V;\kappa) + He(0)$	‡ A classical (binary) collision model was used that includes an adjustable anisotropy factor.
	C	‡	(b) $H_2^+(0;\nu J) + He(0) \to$ $H_2^+(V;\kappa) + He(0)$	(a) $\nu=0$, J=0; three values of $|\kappa|$ were considered. Data are given for the laboratory frame.
	C	‡	(b′) $H_2^+(0;\nu J) + He(0) \to$ $H_2^+(V;\kappa) + He(\Sigma)$	(b,b′) Data for a weighted average of ν,J are given in the lab frame for one value of $|\kappa|$.
	B1	1,2,10	(c) $H_2^+(0;\nu J) + He(0) \to$ $H_2^+(2p\sigma_u;\kappa) + He(0)$	(c) Data for a weighted average of ν,J are given in the center of mass frame for three values of $|\kappa|$.
			(d) the sum of b + c	(d) The sum is for a weighted average of ν,J and is given in the center of mass frame for three values of $|\kappa|$.
			(d) the sum of b′ + c	

TABLE 7.3

Reference	Theory	Approximation	Process	Remarks
205	B1	1,3	$H_2^+(0;2.0) + H^+ \rightarrow$ $H_2^+(2p\sigma_u, 2p\pi_u, 2s\sigma_g, 3d\pi_g; 2.0)$ $+ H^+$	(a) These data are constructed using an empirical scaling law.
	B1	1,3	(a) $H_2^+(0;1\text{-}5) + H^+ \rightarrow$ $H_2^+(2p\sigma_u, 2p\pi_u; 1\text{-}5) + H^+$	(b) $v = 0(1)15$; also give rotational correction factors for small rotational angular momentum J.
	B1	1,3,7	(a) $H_2^+(0;1\text{-}5) + H^+ \rightarrow$ $H_2^+(\Sigma_\sigma'' + \Sigma_\pi'; 1\text{-}5) + H^+$	(c) $v = 0(2)8$; also give rotational correction factors.
	B1	1,4,5	(b) $H_2^+(0;vJ) + H^+ \rightarrow$ $H_2^+(2p\sigma_u; \Sigma) + H^+$	(d) $v = 0(1)15$; give rotational correction factors. Only states high-energy asymptotic results.
	B1	1,4,5,7	(c) $H_2^+(0;vJ) + H^+ \rightarrow$ $H_2^+(\Sigma''; \Sigma) + H^+$	
	C	R	(d) $H_2^+(0;vJ) + H^+ \rightarrow$ $H_2^+(V;\Sigma) + H^+$	
200(c)	*	‡	(a) $H_2(0;vJ) + H^+ \rightarrow$ $H_2(V;\Sigma) + H^+$	* Calculates the B1, C, and sudden approximation cross sections.
	*	‡	(a) $HgH(0;vJ) + H^+ \rightarrow$ $HgH(V;\Sigma) + H^+$	‡ B1 and C results assume a two-body Coulomb interaction potential.
			(a) $NaCl(0;vJ) + H^+ \rightarrow$ $NaCl(V;\Sigma) + H^+$	(a) $v = 0, J = 0$

TABLE 7.3

Reference	Theory	Approximation	Process	Remarks
241(c)	B1*		$H_2(0;1.4) + H^+ \rightarrow$ $2H^+ + H^-(0)$	* This double charge transfer reaction is not based on B1 but involves a product of two B1 single charge transfer amplitudes.
63(c)	B1	1,2,3	$H_2^+(0;1.4\text{-}10) + H(0) \rightarrow$ $H_2^+(2p\sigma_u;1.4\text{-}10) + H(0)$	(a) $R = 1.4(0.4)3.2,4,5,6.8,8.2,9.8;dQ/d\kappa$ in the center of mass frame.
	B1	1,2,10	(a) $H_2^+(0;R) + H(0) \rightarrow$ $H_2^+(2p\sigma_u;\kappa) + H(0)$	(b) $dQ/d\Omega(\kappa)$ in the laboratory frame. Two approximations to the anisotropy of $|\epsilon(\mathbf{K},\kappa)|^2$ are also explored.
	B1	1,2,10	(b) $H_2^+(0;\overline{R}) + H(0) \rightarrow$ $H_2^+(2p\sigma_u;\kappa) + H(0)$	
217	B1	1,2,4,5	$H_2^+(0;\nu J) + H^+ \rightarrow$ $H_2^+(2p\sigma_u;\Sigma) + H^+$	$\nu = 0(1)18; J = 0$
	B1	1,2,3,8,9	$H_2^+(0;2.0) + H(0) \rightarrow$ $H_2^+(2p\sigma_u;\Sigma) + H(\Sigma)$	
	B1	1,2,3,8	$H_2^+(0;2.0) + H(0) \rightarrow$ $H_2^+(2p\sigma_u;\Sigma) + H(\Sigma)$	
	B1	1,2,4,5	$H_2^+(0;\nu J) + H(0) \rightarrow$ $H_2^+(2p\sigma_u;\Sigma) + H(0)$	
	B1	1,2,4,5,8	$H_2^+(0;\nu J) + H(0) \rightarrow$ $H_2^+(2p\sigma_u;\Sigma) + H(\Sigma)$	

161

TABLE 7.3

Reference	Theory	Approximation	Process	Remarks
37(c)	C	B*	$H_2^+(0;\nu J) + X(0) \rightarrow$ $H_2^+(D;\Sigma) + X(0)$	X = 2H, He, 2N, Ar
	C	G,B*	$H_2^+(0;\nu J) + X(0) \rightarrow$ $H_2^+(D;\Sigma) + X(\Sigma')$	$\nu = 0(1)18; J = 0$
	C	R	$H_2^+(0;\nu J) + X(0) \rightarrow$ $H_2^+(V;\Sigma) + X(\Sigma)$	* These results are described as B1. They are classified as classical here since the Born matrix element for H_2^+ was set equal to unity. See the discussion in Sect. 7.3.
	C	B*	$H_2^+(0;\nu J) + X(0) \rightarrow$ $H_2^+(I;\Sigma) + X(0)$	
	C	G,B*	$H_2^+(0;\nu J) + X(0) \rightarrow$ $H_2^+(I;\Sigma) + X(\Sigma')$	
235(c)	†		(a) $H_2^+(0;\nu J) + H(0) \rightarrow$ $H_2^+(2p\sigma_u;\Sigma) + H(0)$	† This calculation is based on the "virtual photon" or Fermi-Williams model.
	†		(b) $H_2^+(0;\nu J) + H(0) \rightarrow$ $H_2^+(2p\sigma_u;\Sigma) + H(\Sigma)$	(a) $\nu = 0, 5, 8; J = 0$ (b) $\nu = 0; J = 0$
236(c)	B1	1,2,4,5	(a) $H_2^+(0;\nu J) + He(0) \rightarrow$ $H_2^+(2p\sigma_u;\Sigma) + He(0)$	$\nu = 0; J = 0$
	B1	1,2,4,6	(a) $H_2^+(0;\nu J) + He(0) \rightarrow$ $H_2^+(V;\Sigma) + He(0)$	(a) Only the sum of these two cross sections is given in the high energy limit. Dependences of these cross sections on the initial vibrational state of H_2^+ is explored.

TABLE 7.3

Reference	Theory	Approximation	Process	Remarks
40(c)	B1	1,2,10	(a) $H_2^+(0;\overline{R}) + H(0) \rightarrow$ $H_2^+(2p\sigma_u;\kappa) + H(0)$	(a) Give $dQ/d\kappa$ in the laboratory frame for scattering angles $\theta = 0°, 0.45°, 1°$, and all observable dissociation velocities.
	B1	1,2,10	(b) $H_2^+(0;\overline{R}) + H(0) \rightarrow$ $H_2^+(2p\pi_u;\kappa) + H(0)$	(b) $dQ/d\kappa$ in the laboratory frame for $\theta = 0°$ and all observable dissociation velocities. Also assume $|\epsilon(\mathbf{K},\kappa)|^2$ is isotopic with respect to κ.
212(c)	B1	1,2,3	$O_2(0;2.3) + H^+ \rightarrow$ $O_2(^3\Sigma_u^-;2.3) + H^+$	* Uses integrated generalized oscillator strengths from electron-O_2 experiment.
	B1	1,4*	$O_2(0;\nu J) + H^+ \rightarrow$ $O_2(^3\Sigma_u^-;\Sigma) + H^+$	
234(c)	B1	1,2,4,5,8	$H_2^+(0;\nu J) + H_2(0;1.4) \rightarrow$ $H_2^+(2p\sigma_u;\Sigma) + H_2(\Sigma;1.4)$	Data are only given for a particular average over initial vibrational state population with $J = 0$. The possibility of simultaneous vibrational excitation of H_2 was ignored.
	B1	1,2,4,5,8	$H_2^+(0;\nu J) + H_2(0;1.4) \rightarrow$ $H_2^+(2p\pi_u;\Sigma) + H_2(\Sigma;1.4)$	
	B1	1,2,4,5,7,8	$H_2^+(0;\nu J) + H_2(0;1.4) \rightarrow$ $H_2^+(\Sigma';\Sigma) + H_2(\Sigma;1.4)$	
231	B1	1,4,5	$H_2^+(0;\nu J) + H^+ \rightarrow$ $H_2^+(2p\sigma_u;\Sigma) + H^+$	$\nu = 0(1)18; J = 0$
	B1	1,4,5	$H_2^+(0;\nu J) + H^+ \rightarrow$ $H_2^+(2p\pi_u;\Sigma) + H^+$	

TABLE 7.3

Reference	Theory	Approximation	Process	Remarks
213(c)	B1	1,4,5,7	$H_2^+(0;\nu J) + H^+ \rightarrow$ $H_2^+(\Sigma'';\Sigma) + H^+$	(a) Only an average over a particular distribution of ν, J=0, is given.
	B1	1,4,5	(a) $H_2^+(0;\nu J) + He(0) \rightarrow$ $H_2^+(2p\sigma_u+2p\pi_u+2s\sigma_g;\Sigma)$ $+ He(0)$	
	B1	1,4,5	(a) $H_2^+(0;\nu J) + He(0) \rightarrow$ $H_2^+(2p\sigma_u+2p\pi_u+2s\sigma_g;\Sigma)$ $+ He(I)$	
	B1	1,4,5	(a) $H_2^+(0;\nu J) + He(0) \rightarrow$ $H_2^+(2p\sigma_u+2p\pi_u+2s\sigma_g;\Sigma)$ $+ He[2^1S+2^1P+3^1P+]$	
	B1	1,4,5,7,8,9	(a) $H_2^+(0;\nu J) + He(0) \rightarrow$ $H_2^+(\Sigma';\Sigma) + He(\Sigma)$	
	B1	1,4,5,8,9	(a) $H_2^+(0;\nu J) + He(0) \rightarrow$ $H_2^+(2p\sigma_u+2p\pi_u+2s\sigma_g;\Sigma)$ $+ He(\Sigma')$	
219	B1	1,2,3	(a) $H_2^+(0;R) + Ar(0) \rightarrow$ $H_2^+(2p\sigma_u,2s\sigma_g,2p\pi_g,2p\pi_u;R)$ $+ Ar(0)$	(a) $1.4 \leqslant R \leqslant 6.0$

TABLE 7.3

Reference	Theory	Approximation	Process	Remarks
	B1	1,2,10	(b) $H_2^+(0;R) + Ar(0) \rightarrow$ $H_2(2p\sigma_u, 2s\sigma_g, 2p\pi_u; \varkappa) + Ar(0)$	(b) Graphs and tables of $dQ/d\varkappa$ in center of mass coordinates. Additional data are available from the authors.
	B1	1,2,10	(b) $H_2^+(0;R) + He(0) \rightarrow$ $H_2^+(2p\sigma_u; \varkappa) + He(2^1P)$	
111(c)	B1	1,2,10	(a) $H_2^+(0;R) + H(0) \rightarrow$ $H_2^+(2p\sigma_u; \varkappa) + H(0)$	$R = 1.0(0.2)6.0$
	B1	1,2,8,9,10	(a) $H_2^+(0;R) + H(0) \rightarrow$ $H_2^+(2p\sigma_u; \varkappa) + H(\Sigma)$	(a) $dQ/d\varkappa$ for center of mass dissociation angles equal to 0° and 90° and collision energies equal to 3.0, 6.0, 10.2, 20.4 keV.
	B1	1,2,10	$H_2^+(0;\overline{R}) + H(0) \rightarrow$ $H_2^+(2p\sigma_u; \varkappa) + H(0)$	
	B1	1,2,8,9,10	$H_2^+(0;\overline{R}) + H(0) \rightarrow$ $H_2^+(2p\sigma_u; \varkappa) + H(\Sigma)$	
211(c)	B1	1,4,5	(a) $H_2^+(0;\nu J) + H^+ \rightarrow$ $H_2^+(2p\sigma_u; \Sigma) + H^+$	(a) $\nu = 0(1)19; J = 0$
	B1	1,4	(b) $H_2^+(0;\nu J) + H^+ \rightarrow$ $H_2^+(2p\sigma_u; \Sigma) + H^+$	These are the only scattering data for $\nu = 19$. (b) $\nu = 15(1)19; J = 0$

7.5 DISCUSSION

In the following discussion a number of qualitative remarks concerning the theoretical results are given to augment the data cited in Tables 7.2 and 7.3. It is hoped that these remarks in some way summarize the situation as it stands today.

No detailed comparison in the form of graphs or tables between theory and experiment are given. Comparisons of various theoretical data are restricted in a similar fashion. Generally speaking, the papers provide this service when possible. The papers providing such information are identified in Table 7.3.

7.5.1 H_2^+—Structureless Charged Particle Collisions

The simplest molecule and the one that has received by far the most theoretical and experimental attention is the hydrogen molecule ion. This is a one-electron molecule, hence most of the required structural information is available or can be obtained by simple procedures. The usual experimental situation is one in which H_2^+ may very likely be in an excited vibrational state prior to the collision. The theory must include this possibility if meaningful comparison with experiment is contemplated. Nothing is known concerning the population of rotational states which would prevail in the typical experiment prior to the collision. Since electron ionization experiments seem not to disturb the initial rotational state population,[228] it has been customary to ignore the possibility of excitations in these degrees of freedom when dealing with H_2^+ formed from the ionization of H_2.

Theory has shown the total cross section for exciting the $1s\sigma_g - 2p\sigma_u$ transition by a structureless charged particle to be very dependent on the initial vibrational state of H_2^+. The cross section for dissociation of the highest vibrational state in the $1s\sigma_g$ orbital is over two orders of magnitude larger than that for the lowest vibrational state; see, for example, ref. 211. Other transitions in H_2^+ are not as dramatically dependent on the initial vibrational state,[205] but one must take vibration into account if highly accurate cross sections are desired. An inspection of Table 7.2 shows that many of the important transitions from the $1s\sigma_g$ orbital of H_2^+ have received some attention. Of practical interest, the ionization of H_2^+ is one transition for which more quantitative data would be useful. A classical calculation,[229] a $B1$ estimate,[230] and a $B1$ upper bound[231] are available. The $B1$ estimate given in ref. 230 is qualitative in nature but proposes an interesting technique, and also a reasonable technique for ionization, for evaluating the effect of vibrational motion. This technique consists of constructing an interpolation formula to connect the ionization cross sections for the combined atom limit and the separated atom limit.

As can be seen from Table 7.2, there have been a number of duplicate studies of the bare charge-H_2^+ system. In general, the agreement between the $B1$ results of various investigators is very good or can be rationalized in terms of the approximations outlined in Section 7.3. The one exception is the data of ref. 224. It has been postulated[211] that this difference may be due to the use of quasiclassical vibrational wave functions in ref. 224. In addition, a certain term appearing in the cross section given by ref. 224, see Eq. 10 of that reference, is claimed to be independent of the initial vibrational state; this conclusion is in conflict with the results of ref. 211.

One study[205] provides information concerning the effects of the initial rotational state on a number of transitions in H_2^+. These data are summarized by formulae resulting from a first-order perturbation treatment of the rotational degrees of freedom. The various cross sections were not observed to be extremely sensitive to the initial rotational state. Hence, unless there is reason to expect unusual initial populations of rotational states, or if there is interest in initial vibrational states near the dissociation limit, these degrees of freedom probably do not cause a prediction based on theoretical data for $J = 0$ to be seriously in error.

Both the $B1$ and classical approximations have been applied to structureless charged particle—H_2^+ collisions. Comparison of these two theories for this specific system can add little to the general information already available.[198] With respect to specific types of collisions, classical arguments have been used to study vibrational dissociation while full $B1$ data for this type of collision are not yet available. In view of the criticisms of classical theory applied to high-energy collisions given in Section 7.3, it seems desirable to have an accurate $B1$ study of dissociation for comparison's sake. An equivalence of classical, $B1$, and sudden approximations has been claimed[200] for vibrational dissociation. The proof relies, however, on some auxiliary approximations (see footnote on page 140), the most restrictive of which seems to be the use of the binary collision model. The results of ref. 200 may be accurate in many situations, but possible limitations on their results are not yet clearly established.

In the context of this discussion, there seems little doubt that $B1$ is preferable to classical theories for specific electronic excitations. The $B1$ theory suffers from the fact that the collision is between charged particles. Hence Coulomb-Born theory should really be used in place of $B1$ theory.[205] At high collision energies, where $B1$ can make some claim for accuracy, one might expect the Coulomb distortion effect to be of minor importance. Again, the statement must be made that quantitative understanding of each particular point has yet to be achieved.

The cross section for the dissociation of H_2^+ by electrons has been measured.[232,233] Since it was argued above that $B1$ theory is applicable to a collision with an arbitrary structureless-charged particle, provided we ignore the exchange channels available in electron collisions, it is appropriate to discuss here the comparison of theory and experiment for this case. The experimentalists[232,233] assumed a particular distribution in initial H_2^+ vibrational states and, using available $B1$ data, constructed the theoretical prediction of the $e^- - H_2^+$ dissociation cross section. The theoretical data were sufficient to give a reasonable estimate, except that it was necessary to use an estimate based on classical arguments to obtain part of the contributions from ionization events. The experimentalists were able to give, in this author's opinion, a rather accurate estimate of the $B1$ prediction of the total dissociation cross section, *provided* the assumed initial vibrational state population was correct and the initial rotational modes were not highly excited. The agreement between experiment and theory was typical of that found for $B1$ predictions. The marked effects due to the initial vibrational state population were confirmed by this comparison. In general, the magnitudes of experiment and theory agreed well; probably fortuitously well for collision energies less than 100 eV. The most disturbing aspect of the comparison concerned the slope, at high collision energies, of a plot of EQ versus log E, where E is the collision energy and Q is the total dissociation cross section. The theoretical slope was almost 30% lower than experiment. There is, as yet, no explanation for this discrepancy, since the slope is a high collision energy property of the cross section, and the known objections to $B1$ theory are not believed to have an effect on this prediction. The dissociation of H_2^+ by heavy charged particles, such as H^+, has not yet been measured. The $H^+ - H_2^+$ system should be quite similar to the $e^- - H_2^+$ system. The main difference would occur at relatively low collision energies. Vibrational dissociation by protons would be more important than it is for electrons. Also, charge transfer is possible for the proton case where electron exchange channels are allowed for the electron case. (Both vibrational dissociation and electron exchange were ignored in the $e^- - H_2^+$ case.)

7.5.2 H_2^+—Neutral Particle Collisions

At high energies where $B1$ should work, it is known that an electronic excitation caused by collision with a neutral particle is quite different from collision with a charged particle. The energy dependence of the total cross section is proportional to v_i^{-2} for large v_i in the neutral case and $v_i^{-2} \, lnv_i$ in the charged case if the transition is dipole-allowed. More important, excitation of internal degrees of freedom of the neutral collision partner must now be taken into account. See ref. 201 for a discussion of H-H inelastic collisions. The usual experimental situation is one in which only the probability for producing a particular dissociation fragment as a function of collision energy is measured. Hence the experimental cross section includes all excitation modes of the molecule that can lead to the particular fragment of interest and all energetically possible excitation modes of the collision partner. This has several consequences. First, the v_i^{-2} behavior occurs at higher energies than one might expect; for example, the proton production cross section for H_2^+ on fairly light targets such as He does not show a v_i^{-2} dependence until the collision energy is in excess of 1 MeV. Also, the proton production cross section for collisions with H_2 and He is relatively flat between 50 and 500 keV. It has been shown from $B1$ calculations[213,234] that the "flat" portion of the curve is due to a combination of a decreasing cross section associated with dissociation accompanied by no excitation of the molecule's collision partner and the increasing cross section corresponding to dissociation accompanied by excitation (ionization) of the collision partner as the energy increases.

Another difference between dissociation of H_2^+ on charged particles and neutral particles is the decreased dependence of the dissociation cross section on the initial vibrational states of H_2^+, except at relatively low collision energies in the case of neutral collisions. First Born studies of the $1s\sigma_g - 2p\sigma_u$ transition in H_2^+-neutral collisions show a very marked dependence on initial vibrational states below 10 keV with decreasing dependence to about 50 keV and fairly constant dependence as the collision energy increases. As one adds other electronic excitation events contributing to H_2^+ dissociation, the dependence on initial vibrational state is further decreased as the energy increases. Unfortunately, $B1$ theory probably cannot be trusted below several hundred keV. By ignoring this, it can be inferred that changing the initial vibrational population has minimum effect on this dissociation cross section at high collision energies, greater than 1 MeV, with this effect becoming more pronounced as the collision energy decreases. Quantitative experimental studies of the dependence of the dissociation cross section on different initial vibrational state populations have not been performed. However, this qualitative theoretical prediction is consistent with most of the experimental data on ion source effects discussed in Chapter 6.

Comparison of theoretical data for dissociation of H_2^+ on collision with several neutral particles is possible. The classical Ru, classical G, and $B1$ theories have been utilized to study H_2^+ collisions with H, H_2, and He. (The classical cross section for collision with H_2 is assumed to be twice the classical cross section for collision with H.) The classical theories have only been used to predict these cross sections at high collision energies, that is, the coefficient of v_i^{-2}. The $B1$ data are available from collision energies from $\sim 10^4$ to more than 10^7 eV.

Comparison of all available theoretical dissociation data and the experimental dissociation data for the H_2 collision partner is presented in ref. 234. Reference 213 contains the same information for the He collision partner. Some general comments can be made on the basis of these comparisons.

The Ru calculation, which ignored the vibrational degree of freedom,[196] is much smaller than the Ru calculation including vibrational effects,[37] as expected. The Ru data are exceeded by the G data.[37] The greater G cross section is undoubtedly explained (mentioned in Section 7.3) by the fact that the G theory allows for the velocity distribution of the "active" particle, whereas the Ru theory does not. From the abstract point of view, the G theory seems preferable to the Ru theory. However, in all cases the Ru data were in better agreement with experiment than the G data.[37] The criticisms, presented in Section 7.3, of the application of classical theories to high-energy collisions are somewhat reinforced by this observation.

A calculation based on an adaptation of the Fermi-Williams model,[235] or "virtual photon exchange" method, has also been used to study the system under consideration. The comparison of this approach with $B1$ theory, presented in ref. 235, shows that the two theories are not even qualitatively in agreement. This is true particularly for the case of simultaneous excitation of the two collision partners. Until further justification for this approach is found, the conclusion that the Fermi-Williams model lacks much promise seems appropriate.

The $B1$ data for dissociation of H_2^+ by H_2[234] and He[213,236] are based on a number of approximations, which are listed in Table 7.3. The effect of these approximations cannot be evaluated quantitatively, but the data of ref. 213 should be of higher quality than that of ref. 234. This observation rests on the fact that ref. 213 used experimentally determined matrix elements for the He target, which appear to be of high quality, whereas ref. 234 used approximate H_2 wave functions in addition to being forced to rely more on closure arguments. The authors of ref. 234 also failed to point out that events resulting in the dissociation of H_2^+, accompanied by vibrational excitation of H_2, were ignored. These events may not be important at high collision energies, but quantitative proof would be more reassuring.

Reference 236 gives the $B1$ cross section in the high-energy limit for two dissociation processes; see Table 7.3 for details. It seems that the use of Approximation 6 in calculating the vibrational dissociation results in a cross section much too large for this event.

The $B1$ predictions[213,234] are uniformly smaller than either of the classical calculations discussed above.[37] Since both the $B1$ theory and the additional approximations become less objectionable in the high collision energy range, we are again led to a difficulty with classical theories. In general, the $B1$ predictions are only slightly smaller than the Ru predictions, and both agree reasonably well with experiment; that is, $B1$ is about 30% or less lower than experiment for the various dissociation mechanisms. The high collision energy data of ref. 236 for He agree fairly well with experiment. This is presumably fortuitous and is due to cancellation of the large positive error expected from the estimation of the vibrational dissociation cross section by the neglect of cross sections which correspond to simultaneous dissociation of H_2^+ and excitation of the helium atom.

For collision energies between 10^4 and 10^6 eV, where the cross section does not display its asymptotic energy dependence, the agreement between $B1$ theory and experiment is rather good for dissociation events by H_2 that exclude the ionization of H_2^+.[234] The same comparison for the He case[213] indicates that theory is around 30% lower than experiment for the larger energies in this range and with an error increasing to almost 80% at 10^5 eV. The discrepancy at lower collision energies in the He case could be attributed to the failure of $B1$ theory. However, the experimental and theoretical evidence does not provide an unequivocal explanation of the rather better agreement in the H_2 case than in the He case. Note, from the preceding discussion, that the He calculation is presumed to be the more accurate of the two studies.

7.5.3 Angular and Velocity Distributions of H_2^+ Dissociation Fragments

Measurements of the velocity and/or angular distribution of dissociation fragments resulting from the dissociation of H_2^+ have stimulated theoretical interest in this topic. See, for example, the literature cited in ref. 219. The possibility of anisotropy with respect to the angle between the internuclear axis and the momentum transfer vector in the cross sections for certain inelastic events was recognized some time ago.[237] Qualitative experimental verification of this observation for $e^- - H_2$ collisions[238] and additional theoretical data for $e^- - H_2^+$ angular distributions is available.[227] As yet there is no possibility of comparing different theoretical predictions or theory with experiment for the bare charge, or electron, case, so the following discussion concentrates on dissociation by neutral particle collisions.

A paper by Kerner,[239] which is concerned with $e^- - H_2^+$ collisions, is highly recommended to the reader for a good exposition of the difficulties involved in studying $B1$ theory for the angular and velocity distributions of dissociation fragments. Reference 219 contains a general, and more complete, statement of the problem, along with the reduction of the general $B1$ theory to the reflection approximation formula. It should be remembered that the analysis of $\epsilon(\mathbf{K}, \mathbf{\kappa})$, Eqs. 7.21 and 7.22, is applicable to both the structureless charged particle and neutral collision partner cases because of the form of the matrix element defined by Eq. 7.16. As stated in Section 7.3, all studies containing sufficient data for meaningful comparison with experiment are based on the reflection model. Reference to the discussion in Section 7.3, especially Approximation 10, gives a qualitative perspective for judging the data discussed below.

Reference 63 was the first paper to present both theory and experiment for the topic under consideration. The experiment gave the angular distribution of H_2^+ in the laboratory frame, resulting from the dissociation of H_2^+ on collision with H_2 at 10^4 eV. Theory was assumed to be twice the cross section for the $1s\sigma_g - 2p\sigma_u$ transition excited in $H_2^+ - H$ collisions. The theoretical center-of-mass cross section, differential in $d\mathbf{\kappa}$, was folded in order to give the appropriate lab angular distribution. This folding ignored the momentum transfer between the centers of mass of the two colliding partners. Subsequent studies[240] indicate that the momentum transfer effects can be very important for some collisions. Other events, such as vibrational dissociation, that may be important for collisions at 10^4 eV, were ignored. Also, the use of $B1$ theory for heavy particle collisions at 10^4 eV is open to question. Despite these objections, theory and experiment, when normalized, agreed remarkably well. The prediction near-zero lab angle was badly in error, but the prediction of prominent "wings" on the angular distribution was verified by experiment. A similar measurement was made of the corresponding angular distribution of H. This angular distribution was qualitatively different from the H^+ distribution and was not considered theoretically because of the lack of theoretical data for the dissociative-charge-exchange events. A measurement of proton velocity distribution as a function of laboratory angle was found to agree qualitatively with a theoretical calculation based on the same premises utilized in ref. 63.[40]

Additional experimental studies make it apparent that the $B1$ theory alone for $1s\sigma_g - 2p\sigma_u$ transitions in H_2^+ is not sufficient to explain quantitatively the observed angular and velocity distributions. (See the discussion and appropriate references cited in ref. 240.) The experimental data are for 10^3 to 10^4 eV H_2^+ colliding with neutral targets. This collision energy range makes use of $B1$ suspect; however, because of the complexity of the collision system, data from more accurate theories will probably not be available soon. Ignoring this basic difficulty, the following speculations are made concerning the available data. Most of the mechanisms discussed below have been considered, to some degree, in experimental papers.[43,111]

Simultaneous excitations of the target and molecule have been studied and found to have a large effect on the predicted velocity and/or angular distributions. Reference 111 considers the possibility of simultaneous excitation events in an approximate way (see Table 7.3), and obtains some justification for their procedures by comparison with experiment. This type of event is known, however, to give a small contribution to the total dissociation cross section in the 10^3 to 10^4 eV range. It seems reasonable to assume that the simultaneous excitation events usually do not significantly contribute to the observed differential cross sections. One exception may be the cross section corresponding to large relative dissociation velocity. These events require a somewhat larger momentum transfer, hence excitation of the neutral collision partner may be more likely. The argument against simultaneous excitation events in this energy range is considered in more detail in ref. 219.

Dissociation events resulting from higher excited electronic states of H_2^+ were also ignored in the comparisons cited above. Theoretical arguments for the importance of these events have been advanced[40,219] and deductions from experiment[44] give supporting evidence. A firm statement cannot be made on the basis of available data, but it would appear likely that consideration of higher electronic states is necessary for quantitative theoretical predictions in the 10^3 to 10^4 eV energy range. They certainly become important for higher collision energies.

One of the most glaring difficulties with the theoretical predictions was the inability to predict a noticeable and narrow peak in the cross section around 90° in the center-of-mass frame, for example, see ref. 43. This peak has been clearly, albeit qualitatively, explained as being due to vibrational dissociation.[240]

The addition of the vibrational dissociation information completes a good qualitative understanding of the angular and/or velocity distribution of dissociation fragments from H_2^+. However, it can be anticipated that further quantitative study of these events will prove that the discussion above has only uncovered a small portion of the interesting science.

7.6　CONCLUDING REMARKS

The preceding discussion of the H_2^+ system includes references to the majority of data cited in Tables 7.2 and 7.3. Although discussion of the remaining studies is required for completeness, this is not done since there would be little opportunity for comparative evaluation.

This review concludes that $B1$ theory can sometimes explain quantitatively, and usually explain qualitatively, the properties of dissociative collisions between heavy particles. The main difficulties arise from dissociation by vibrational events and by events involving charge transfer. First Born theory has not been adequately explored for vibrational dissociation events, but one does not expect $B1$ to be completely reliable for these predominantly close encounters. Only one theoretical study of a dissociative charge-exchange collision has been attempted so far.[241] Although ref. 241 uses an adaptation of $B1$ theory to double charge-exchange and obtains semiquantitative agreement with experiment, it is too early to evaluate $B1$ theory as applied to a general dissociative charge-exchange collision. It should be remembered that difficulties in applying $B1$ theory to rearrangement in collisions between simpler systems are well documented.[242] The opinion has been taken that classical theories are suspect when applied to very high-energy collisions. However, the simple classical approach described here, and the many possible refinements,[198] may yet prove useful for dissociation at intermediate collision energies as it has in other collisions.

Dissatisfaction with $B1$ and classical theories is bound to grow as time goes on. Indeed, a study of molecular collisions by a modification of $B1$ theory, based on the perturbed stationary state-impact parameter formalism, has recently been undertaken.[243] The difficulties associated with an improved theory are great. However, the study contained in ref. 243 already provides insight into the vibrational excitation mechanism not given by the $B1$ theory.

APPENDIX I

The following table summarizes the results of our correspondence with the authors of papers on dissociation. Symbols at column headings are defined as follows:

I A specific inquiry concerning this reference was directed to the author by letter.

S Reference was supplied by author in response to an inquiry concerning some other work.

R Reply was received.

TABLE A.1

Ref. No.	I	S	R	Comments
6	X		X	
7	X			
8	X		X	The gas in the analyzer was air except for ~1.6 × 10⁻⁶ torr background pressure from ion source (Kupriyanov 5-16-68).
9	X		X	Figs. 4, 5 are for 2 keV ion energy; Fig. 9 is for 2 keV ion energy. The word "total" is wrong on Fig. 11 ordinate scale. (McGowan 4-30-69)
10	X		X	The word "total" is misleading on the cross section scales. (McGowan 4-30-69)
11	X		X	Figs. 1 and 6 are erroneously labeled total cross section. Cross sections are lower limits. (McGowan 4-30-69)
12	X		X	In Fig. 3 energy of ions is 2 keV. Cross sections are lower limits. (McGowan 4-30-69)
15	X			
17		X		Cross section for $CO^{+2} + CO \rightarrow 0^+$ should read 1.4×10^{-16} cm^2. There is a sizeable systematic error in the calibration procedure. The assumption that the length of the effective collision volume is not dependent on the radii of the ions is not justifiable. (Harris 8-21-68)
18	X			
21	X		X	Some of data in Ref. 21 were presented earlier in 249. (Sweetman 3-18-68)
22	X			Discovered this after we wrote to Kupriyanov. (McClure)

TABLE A.1

Ref. No.	I	S	R	Comments
23	X		X	Reply received from Solovev.
24	X		X	The data in 24 except those in Fig. 6 are given more precisely in 100. (Solovev 11-1-67)
25	X		X	
26	X			
27	X		X	
28	X		X	
29	X			
30	X			
31	X		X	Accuracy of data in 31 difficult to assess other than general category of 10% or so. (Rourke 11-21-68). In Table IV He was the target gas in all cases. Some of the molecular ion species were also checked with other gases to assure the independence of the T values with respect to the target gas. The energy of the projectile ion over the whole table was tested from 17.5 to 30 keV for singly charged ions and from 30 to 70 keV for doubly charged ions to check the independence of T value with respect to projectile energy. The best nominal projectile energy value would be about 25 keV. (Rourke 2-25-69)
35	X			

Ref. No.	I	S	R	Comments
36	X		X	This is the whole of work on dissociation published under my name. See Ref. 35 for other work of this laboratory. (Fremlin 5-9-68)
37	X		X	Ref. 37 includes all my work on H_2^+ dissociation. Best estimate of experimental uncertainties given in paper. (Berkner 11-28-67)
38	X		X	Results in 38 are not too accurate as apparatus was not designed for the purpose. Used rough transformation to c.m. coordinates, hence central peak is anomalous. (Caudano 11-26-67)
39	X		X	
40	X		X	A number of references were suggested including 237 and 224. Also attention was drawn to experimental work in preprint form concerning the dissociation of $(HeH)^+$ and to a theoretical preprint by A. Russek (see remarks for Ref. 235). (Los 7-22-69)
41	X		X	Results in Ref. 41 are incorrect and are superseded by Refs. 42 and 43. (Gibson 3-27-68)
42	X		X	The data in Ref. 42 are duplicated in Ref. 43. (McClure)
43	X		X	Data reproducibility was ±5%. (Gibson 3-27-68)
44		X		Ref. 44 is more accurate and more complete than Ref. 258. (Vogler 1-3-68)
45	X		X	Data in 45 are corrected version of Ref. 257 data. (Vance 2-12-68)
46	X		X	Ref. 46 employs same methods and accuracy as 21. Not published. (Sweetman 3-18-68)

178

TABLE A.1

Ref. No.	I	S	R	Comments
47	X		X	My only publication on the subject. No revisions known to be necessary. (Riviere 2-28-68)
48	X		X	This is the most complete and precise of my papers on H_2^+ dissociation. (Guidini 10-17-62). Ref. 48 apparently duplicates or supersedes Refs. 253, 254, 255 (McClure)
49	X		X	Reply from H. Spehl coauthor.
50	X		X	
57		X		
58		X		Data Ref. 58 appears to be the same as those on Ref. 127. (McClure)
62		X		Kupriyanov sent reprint with his reply.
63	X		X	
64	X		X	The slit dimensions in Ref. 64 were (mm) $S_1 = 1 \times 7$, $S_2 = 3 \times 10$, $S_3 = 15 \times 10$. In Fig. 6 along the ordinate axis there should be numerals 2 and 4 instead of 1 and 2 (Kupriyanov 5-16-68).
65	X		X	
66	X			

TABLE A.1

Ref. No.	I	S	R	Comments
67	X		X	Total uncertainty is 10% for + ions and 20% for negative ions based on reproducibility. (Ogurtsov 3-4-68)
68	X		X	
69	X		X	
70	X			
71	X		X	Results in Ref. 71 are only abbreviated results of a much fuller report in Ref. 151. (Lindholm 4-29-68)
72		X		
75		X		
76	X		X	
77	X		X	My only publication on dissociation induced by charge exchange. (Galli 10-17-67)
79	X		X	
81	X		X	Accuracy not better than ±20%. This is our only publication on dissociation. (Tiernan 7-9-68)
83	X		X	This Ref. 83 is very much a preliminary report. (Homer 3-5-68)
84	X		X	This Ref. 84 is very much a preliminary report. (Homer 3-5-68)
85	X			

TABLE A.1

Ref. No.	I	S	R	Comments
86		X	X	Williams suggested we include this type of dissociation data for completeness. (Williams 3-18-68)
87	X	X		
88		X		
90		X		Ref. 90 apparently supersedes Ref. 259 (McClure). Accuracy is $\pm40\%$ based on random and calibration errors. (Sinda 2-16-68)
91	X			Thesis. Part of Ref. 91 dissociation data published in Ref. 25. (McClure)
92	X			
93	X		X	Thesis. All dissociation data from Ref. 93 are published in 16. (Harris 8-21-68)
94		X		Thesis. Results partly published in 72, 161, 177, 178. (McClure)
95				Thesis. No letter written.
96	X		X	Thesis. Data partly published in 103. (McClure)
97	X		X	Thesis. Data from Ref. 97 are published in one or more of Refs. 159, 171, 186, 187. (Hertel 5-15-68)
98	X			Thesis. Partly published Ref. 129. (McClure)
99	X			
100	X		X	

TABLE A.1

Ref. No.	I	S	R	Comments
101		X		
102		X		
103	X			
105	X		X	Thesis. Apparently the dissociation data are published in 92. (McClure)
106	X		X	
107	X		X	Data in Ref. 107 are preliminary. (Caudano 11-26-67)
108	X		X	
109	X		X	The maximum angular spread was 1.3 degrees for H_2^+ and H_3^+ ions entering collision chamber. (Chambers 10-9-67)
112	X			
113	X		X	Mean free paths should be labeled 1_1, 1_2, 1_3, not i_1, i_2, i_3. (Goldring 2-5-68)
114	X		X	In Figs. 1 and 2, 20 keV should read 18 keV. (Steyaert 6-12-69)
115	X		X	
117	X			
118	X		X	
119		X		
120	X		X	Data on D_2^+ collisions with He and O_2 were never published and cannot be found.

TABLE A.1

Ref. No.	I	S	R	Comments
125		X		
126	X		X	Data on D_2^+ collisions with Ar and N_2 were published in Ref. 79. (Bailey 3-13-68) (Champion 9-21-67) The dimensions of the collimating slits were as follows: (mm)
				S1 S2 S3 — Ethylene 1×7, 2×10, 1.5×10; Propane 0.5×7, 1.5×10, 1.5×10; Other gases 1×7, 3×10, 14×10 (Kupriyanov 5-16-68)
127	X		X	The slit dimensions in Ref. 127 were (mm) $S_1 = 0.5 \times 7$, $S_2 = 1.5 \times 10$, $S_3 = 1.5 \times 10$. (Kupriyanov 5-16-68)
129	X			
130	X		X	(Reply from Lindholm – no comment)
131	X			
132	X		X	
136	X		X	Units of cross section is cm^{-1}. (Lindholm 4-29-68)
137	X			
138	X		X	I suspect errors in Ref. 138 due to $\eta < 1$ are not greater than 30%, but errors of 2 to 3 due to $\eta < 1$ are not impossible. (Maier 2-2-68)

TABLE A.1

Ref. No.	I	S	R	Comments
139	X		X	
141	X			
142		X		Submitted to Planetary & Space Sciences.
143		X		
144	X			
145	X			
146			X	Inquired about Ref. 145 by same author – no reply.
147	X		X	
148	X		X	The experiment of Ref. 148 was carried out under the same conditions as Ref. 64 (Kupriyanov 5-16-68). See notes on Ref. 64. (McClure)
149		X		
150	X		X	The slit dimensions in Ref. 150 were (mm) $S_1 = 1 \times 7$, $S_2 = 3 \times 10$, $S_3 = 14 \times 10$. (Kupriyanov 5-16-68)
151	X		X	
153	X		X	
154	X			

TABLE A.1

Ref. No.	I	S	R	Comments
155	X			
156	X		X	
158		X		
159	X		X	
160	X			
161		X		
162	X			
163	X			
164	X		X	
165	X			
166	X			
167		X		
168	X			
169		X		
170	X		X	
171	X		X	
172	X			

185

TABLE A.1

Ref. No.	I	S	R	Comments
173	X		X	
175	X			
176	X		X	
177	X		X	
178	X		X	
179	X		X	Reply from Lindholm (no comment).
180	X		X	
181	X		X	
182	X		X	
183	X		X	Reply from Lindholm (no comment).
185	X			
186		X		
187	X		X	Similar curves to those for Ar, but for other gases, are in Hertel's thesis: Ref. 97. (Hertel 2-14-68)
196	X			
200	X			
205	X		X	

TABLE A.1

Ref. No.	1	S	R	Comments
214		X		See remarks for Ref. 245.
224		X		See remarks for Ref. 40.
227		X		See remarks for Ref. 236.
229	X		X	
230	X		X	The author emphasized the point that the motivation for this work was toward a qualitative understanding. (Gerjuoy 9-4-69)
235	X		X	The right hand side of the second and third equations from the bottom of page 191 should be multiplied by 2π. Also the existence of a preprint by A. Russek was mentioned. (Durup 9-18-69)
236	X		X	This problem is receiving continued attention with emphasis on theories other than first Born. The inclusion of Ref. 227 was suggested. (Farina 9-25-69)
237		X		See remarks for Ref. 40.
241	X		X	
244	X		X	
245,246	X		X	This author suggested the addition of Ref. 246. (Bauer 6-30-69)
247	X		X	The contents of this paper have been redone and are available in report form from the author. The importance of reading Ref. 248 in conjunction with Ref. 247 was mentioned (Breen – not dated but posted after 7-1-69).

TABLE A.1

Ref. No.	1	S	R	Comments
248	X		X	See Ref. 247.
250	X		X	Reference 250 should be eliminated as the data are in error. (Barnett 9-25-67)
251	X		X	Data presented in 251 are superseded by data in Ref. 37. (Pyle 4-9-68)
252	X		X	It would be preferable to use data from Ref. 76. (Fite 10-16-67)
253	X		X	Superseded by 48 (see note on Ref. 48).
254	X		X	Superseded by 48 (see note on Ref. 48).
255	X		X	Superseded by 48 (see note on Ref. 48).
256				Apparently superseded by Ref. 92 (McClure).
257	X		X	Data in Ref. 257 were not corrected for currents collected by grids (Vance 2-12-68). See Ref. 44 (McClure).
258	X		X	Ref. 44 is more accurate and more complete. (Vogler 1-3-68)
259	X		X	Apparently Ref. 259 is superseded by Ref. 90. (McClure)
260				Data published in Ref. 132. (Solovev 3-4-68)

REFERENCES

1. H. S. W. Massey and E. H. S. Burhop, *Electronic and Ionic Impact Pheno-mena*, Clarendon Press, Oxford (1956).
2. J. B. Hasted, *Physics of Atomic Collisions*, Butterworth and Co. Ltd., London (1964).
3. E. W. McDaniel, *Collision Phenomena in Ionized Gases*, Wiley, New York, (1964), pp. 1–2.
4. B. Bederson and A. L. Fite, *Methods of Experimental Physics*, Vol. 7, Academic Press, New York and London (1968).
5. H. Ewald and H. Hintenberger, AEC-TR-5080 (1962).
6. G. W. McClure, *Phys. Rev.*, **153**, 182 (1967).
7. N. N. Tunitskii, R. M. Smirnova, and M. V. Tikhomirov, *Dokl. Akad. Nauk SSSR*, **101**, 1083 (1955) (English trans.: AEC-TR-2310).
8. S. E. Kupriyanov, M. V. Tikhomirov, V. K. Potapov, and P. I. Karpova, *Zh. Eksp. Teor. Fiz.*, **30**, 569 (1956) [English trans.: *Sov. Phys.-JETP*, **3**, 459 (1956)].
9. Wm. McGowan and L. Kerwin, *Can. J. Phys.*, **41**, 316 (1963).
10. J. W. McGowan and L. Kerwin, *Can. J. Phys.*, **42**, 972 (1964).
11. Wm. McGowan and L. Kerwin, *Can. J. Phys.*, **42**, 2086 (1964).
12. Wm. McGowan and L. Kerwin, *Proc. Phys. Soc. (London)*, **82**, 357 (1963).
13. E. Friedlander, H. Kallmann, W. Lasareff, and B. Rosen, *Z. Physik*, **76**, 60 (1932).
14. J. A. Hipple, R. E. Fox, and E. U. Condon, *Phys. Rev.*, **69**, 347 (1946).
15. V. K. Potapov, *Zh. Fiz. Khim. Akad. Nauk SSSR*, **34**, 444 (1960) [English trans.: *Russ. J. Phys. Chem.*, **34**, 207 (1960)].
16. J. H. Beynon, R. A. Saunders, and A. E. Williams, *Z. Naturforsch.*, **20a**, 180 (1965).
17. H. H. Harris and M. E. Russell, *J. Chem. Phys.*, **47**, 2270 (1967).
18. B. N. Makov, L. I. Elizarov, and E. A. Dmitriev, *Instr. Exper, Tech.*, **2**, 390 (1964).
19. A. Henglein, *Z. Naturforsch.*, **7a**, 165 (1952).
20. A. Henglein and H. Ewald, *Mass Spectroscopy in Physics Research*, NBS Circular 522, 205 (1953).
21. D. R. Sweetman, *Proc. Roy. Soc. (London)* **A256**, 416 (1960).
22. S. E. Krupiyanov, *Prob. Fiz. Khim.*, **3**, 76 (1962).
23. N. V. Fedorenko, *Zh. Tekh. Fiz.*, **24**, 769 (1954) (Translation RJ-540, Association Technical Services, Inc., East Orange, N. J.).
24. N. V. Fedorenko, V. V. Afrosimov, R. N. Il'in, and D. M. Kaminker, *Zh. Eksp. Teor. Fiz.*, **36**, 385 (1959) [English trans.: *Sov. Phys.-JETP*, **9**, 267 (1959)].
25. J. F. Williams and D. N. F. Dunbar, *Phys. Rev.*, **149**, 62 (1966).
26. K. K. Damodaran, *Proc. Roy. Soc. (London)*, **A239**, 382 (1957).
27. G. W. McClure, *Phys. Rev.*, **130**, 1852 (1963).
28. C. F. Barnett and J. A. Ray, *Atomic Collision Processes*, North Holland Publ. Co., Amsterdam (1964), p. 743.
29. L. I. Pivovar, V. M. Tubaev, and M. T. Novikov, *Zh. Eksp. Teor. Fiz.*, **40**, 34 (1961) [English trans.: *Sov. Phys.-JETP*, **13**, 23 (1961)].

30. H. Postma and D. P. Hamblen, Oak Ridge National Laboratory-2966, Oak Ridge, Tenn. (1960).
31. F. M. Rourke, J. C. Sheffield, W. D. Davis, and F. A. White, *J. Chem. Phys.*, **31**, 193 (1959).
32. W. W. Hunt, Jr., and K. E. McGee, *J. Chem. Phys.*, **41**, 2709 (1964).
33. W. W. Hunt, Jr., R. E. Huffman, and K. E. McGee, *Rev. Sci. Instr.*, **35**, 82 (1964).
34. R. E. Ferguson, K. E. McCulloh, and H. M. Rosenstock, *J. Chem. Phys.*, **42**, 100 (1965).
35. K. E. A. Effat, *Proc. Phys. Soc. (London)*, **A65**, 433 (1952).
36. J. H. Fremlin and V. M. Spiers, *Proc. Phys. Soc. (London)*, **A68**, 398 (1955).
37. K. H. Berkner, S. N. Kaplan, R. V. Pyle, and J. W. Stearns, *Phys. Rev.*, **146**, 9 (1966).
38. R. Caudano, J. M. Delfosse, and J. Steyaert, *Ann. Soc. Sci. Bruxelles,* **I 76**, 127 (1962).
39. J. Durup, P. Fournier, and D. Pham, *Int. J. Mass Spectrom. Ion Phys.*, **2**, 311 (1969).
40. D. K. Gibson and J. Los, *Physica,* **35**, 258 (1967).
41. D. K. Gibson, J. Los, and J. Schopman, V International Conference on the Physics of Electronics and Atomic Collisions, Leningrad, USSR, 1967. p. 594.
42. D. K. Gibson, J. Los, and J. Schopman, *Phys. Letters,* **25a**, 634 (1967).
43. D. K. Gibson, J. Los, and J. Schopman, *Physica,* **40**, 385 (1968).
44. M. Vogler and W. Seibt, *Z. Physik,* **210**, 337 (1968).
45. D. W. Vance and T. L. Bailey, *J. Chem. Phys.*, **44**, 486 (1966).
46. D. R. Sweetman, private communication (1965).
47. A. C. Riviere and D. R. Sweetman, *Proc. Phys. Soc. (London)*, **78**, 1215 (1961).
48. J. Guidini, *Compt. Rend.*, **253**, 829 (1961).
49. H. Ropke and H. Spehl, *Nucl. Instr. Meth.*, **17**, 169 (1962).
50. S. E. Kupriyanov, *Zh. Tech. Fiz.*, **34**, 861 (1964) [English trans.: *Sov. Phys.-Tech. Phys.*, **9**, 659 (1964)].
51. C. Ottinger, *Phys. Letters,* **17**, 269 (1965).
52. C. Ottinger, *Z. Naturforsch.*, **20a**, 1229 (1965).
53. M. C. Flowers, *Chem. Commun.*, **11**, 235 (1965).
54. W. Sonneveld, *Rec. Trav. Chim.*, **84**, 45 (1965).
55. A. S. Newton and A. F. Sciamanna, *J. Chem. Phys.*, **40**, 718 (1964).
56. V. H. Dibeler, C. E. Wise, Jr., and F. L. Mohler, *Phys. Rev.*, **71**, 381 (1947).
57. S. E. Kupriyanov and A. A. Perov, *Dokl. Akad. Nauk SSSR.* **149**, 1368 (1963) [English trans.: *Dokl. Phys. Chem.*, **149**, 351 (1963)].
58. S. E. Kupriyanov and A. A. Perov, *Zh. Fiz. Khim. Akad. Nauk SSSR,* **39**, 1640 (1965) [English trans.: *Russ. J. Phys. Chem.*, **39**, 871 (1965)].
59. J. A. Hipple and E. U. Condon, *Phys. Rev.*, **68**, 54 (1945).
60. J. A. Hipple, *Phys. Rev.*, **71**, 594 (1947).
61. H. M. Rosenstock and C. E. Melton, *J. Chem. Phys.*, **26**, 314 (1957).
62. N. N. Tunitskii, S. E. Kupriyanov, and A. A. Perov, *Izvert. Akad. Nauk SSSR, Otd. Khim. Nauk,* **11**, 1945 (1962) [English trans.: *Bul. Acad. Sci. (USSR)*, **7**, 1857 (1962)].
63. G. W. McClure, *Phys. Rev.*, **140A**, 769 (1965).
64. S. E. Kupriyanov, N. N. Tunitskii, and A. A. Perov, *Zh. Tekh. Fiz.*, **33**, 1252 (1963) [English trans.: *Sov. Phys.-Tech. Phys.*, **8**, 932 (1964)].

65. G. W. McClure, *Phys. Rev.*, **134**, A1226 (1964).
66. D. V. Pilipenko and Y. M. Fogel, *Zh. Eksp. Teor. Fiz.*, **48**, 404 (1965) [English trans.: *Sov. Phys.-JETP*, **21**, 266 1965)].
67. G. N. Ogurtsov and I. P. Flaks, *Zh. Tekh. Fiz.*, **36**, 117 1966) [English trans.: *Sov. Phys.-Tech. Phys.*, **11**, 84 1966)].
68. E. S. Solov'ev, R. N. Il'in, V. A. Oparin, and N. V. Fedorenko, *Atomic Collision Processes*, North Holland Publ. Co., Amsterdam (1964), p. 692.
69. A. Schmid, *Z. Physik*, **161**, 550 (1961).
70. G. K. Lavrovskaya, M. I. Markin, and V. L. Tal'Roze, *Kinetica i Kataliz*, **2**, 21 (1961) [English trans.: *Kinetics and Catalysis*, **2**, 18 (1961)].
71. E. Lindholm, *Applied Mass Spectrometry*, Institute of Petroleum, London (1954), p. 191.
72. J. C. Abbe and J. P. Adloff, *Bull. Soc. Chim. France*, 1212 (1964).
73. N. V. Fedorenko and V. V. Afrosimov, *Zh. Tekh. Fiz.*, **26**, 1941 (1956) [English trans.: *Sov. Phys.-Tech. Phys.*, **1**, 1872 (1957)].
74. C. F. Giese and W. B. Maier, *J. Chem. Phys.*, **39**, 739 (1963).
75. R. F. Stebbings, J. A. Rutherford, and B. R. Turner, *Planet. Space Sci.*, **13**, 1125 (1965).
76. R. F. Stebbings, A. C. H. Smith, and H. Ehrhardt, *J. Chem. Phys.*, **39**, 968 (1963).
77. A. Galli, A. Giardini-Guidoni, and G. G. Volpi, *Nuovo Cimento*, **31**, 1145 (1964).
78. R. Browning and H. B. Gilbody, *J. Phys. B* (*Proc. Phys. Soc.*), Ser. 2, **1**, 1149 (1968).
79. R. L. Champion, L. D. Doverspike, and T. L. Bailey, *J. Chem. Phys.*, **45**, 4377 (1966).
80. V. Cermak and Z. Herman, *Nucleonics*, **19**, 106 (1961).
81. J. H. Futrell and T. O. Tiernan, *J. Chem. Phys.*, **39**, 2539 (1963).
82. A. Henglein and G. A. Muccini, *Z. Naturforsch.*, **17a**, 452 (1962).
83. J. B. Homer, R. S. Lehrle, J. C. Robb, M. Takahasi, and D. W. Thomas, *Adv. in Mass Spectry.*, **2**, 503 (1963).
84. J. B. Homer, R. S. Lehrle, J. C. Robb, and D. W. Thomas, *Adv. in Mass Spectry.*, **3**, 415 (1966).
85. I. M. Fogel, R. V. Mitin, V. F. Kozlov, and N. D. Romashko, *Zh. Eksp. Teor. Fiz.*, **35**, 565 (1958) [English trans.: *Sov. Phys.-JETP.*, **8**, 390 (1959)].
86. J. F. Williams, *Phys. Rev.*, **157**, 97 (1967).
87. G. W. McClure, *Phys. Rev.*, **132**, No. 4, 1636 (1963).
88. J. F. Williams, *Phys Rev.*, **150**, 7 (1966).
89. E. W. Thomas, Report No. ORO-2591-35, Georgia Tech. (1968).
90. T. Sinda, C. Manus, and J. Guidini, *C. R. Acad. Sc. Paris*, **264**, 755 (1967).
91. J. F. Williams, thesis, Australian Natl. Univ. (1965).
92. F. P. G. Valckx and P. Verveer, *J. Phys.*, **27**, 480 (1966).
93. H. H. Harris, thesis, Michigan State Univ. (1967).
94. J. C. Abbe, thesis, Faculte des Sciences de l'Universite de Strasbourg (1967).
95. W. Muller-Duysing, thesis, University of Hamburg (1963).
96. M. G. Menendez, thesis, Univ. of Florida (1963).
97. G. R. Hertel, dissertation, Johns Hopkins Univ (1965).
98. K. M. A. Rafaey, thesis, Argonne National Lab Informal Report PHY-1966A. (1966).
99. C. E. Melton and G. F. Wells, *J. Chem. Phys.*, **27**, 1132 (1957).

100. R. N. Il'in, B. I. Kikiani, V. A. Oparin, E. S. Solov'ev, and N. V. Fedorenko, *Zh. Eksp. Teor. Fiz.*, **46**, 1208 (1964) [English trans.: *Sov. Phys.-JETP*, **19**, 817 (1964)].

101. S. E. Kupriyanov and A. A. Perov, *Zh. Fiz. Khim. Akad. Nauk SSSR*, **38**, 2263 (1964) [English trans.: *Russ. J. Phys. Chem.*, **38**, 1222 (1964)].

102. S. E. Kupriyanov and A. A. Perov, *Dokl. Akad. Nauk SSSR*, **158**, 942 (1964) [English trans.: *Dokl. Phys. Chem.*, **158**, 923 (1964)].

103. M. G. Menendez, B. S. Thomas, and T. L. Bailey, *J. Chem. Phys.*, **42**, 802 (1965).

104. R. Caudano, and J. M. Delfosse, *J. Phys.* **B1**, 813 (1968).

105. P. Verveer, CEA-R 2983 (1966).

106. H. Spehl and H. Ropke, *Z. Physik*, **166**, 311 (1962).

107. R. Caudano and J. M. Delfosse, V International Conference on the Physics of Electronics and Atomic Collisions, Leningrad, USSR, p. 590, 1967.

108. S. E. Kupriyanov and V. K. Potapov, *Zh. Eksp. Teor. Fiz.*, **33**, 311 (1957) [English trans.: *Sov. Phys.-JETP*, **6**, 244 (1958)].

109. E. S. Chambers, *Phys. Rev.*, **139**, A1068 (1965).

110. T. F. Moran and D. C. Fullerton, *Chem. Phys. Letters*, **2**, 625 (1968).

111. B. Meierjohann and W. Seibt, *Z. Physik*, **225**, 9 (1969).

112. K. H. Purser, P. H. Rose, N. B. Brooks, R. P. Bastide, and A. B. Wittkower, *Phys. Letters*, **6**, 176 (1963).

113. G. Goldring, D. Kedem, U. Smilansky, and Z. Vager, *Nucl. Instr. Meth.*, **23**, 231 (1963).

114. J. Steyaert, J. Bouchat, and J. Delfosse, *Ann. Soc. Sci. Bruxelles*, **75**, 188 (1961).

115. C. F. Barnett, J. A. Ray, and J. C. Thompson, ORNL-3652, 78 (1964).

116. R. W. Rozett and W. S. Koski, *J. Chem. Phys.*, **49**, No. 6, 2691 (1968).

117. A. P. Irsa and L. Friedman, *J. Chem. Phys.*, **34**, 330 (1961).

118. S. E. Kupriyanov, A. A. Perov, and N. N. Tunitskii, *Zh. Eksp. Teor. Fiz.*, **43**, 1152 (1962) [English trans.: *Sov. Phys.-JETP*, **16**, 1152 (1963)].

119. J. Durup and M. Durup, *J. Chim. Phys.*, **2**, 386 (1967).

120. R. L. Champion, T. L. Bailey, and L. D. Doverspike, *Bull. Am. Phys. Soc.*, **11**, 532 (1966).

121. J. C. Houver, J. Baudon, M. Abignoli, and M. Barat, Int. J. Mass Spectrom. Ion Physics **4**, 137 (1970).

122. J. Schopman and J. Los, *Phys. Letters*, **31A**, 79 (1970).

123. J. Schopman, A. K. Barua, and J. Los, *Phys. Leters*, **29A**, p. 112 (1969).

124. J. Schopman and J. Los, *Physica* **48**, 190 (1970).

125. V. M. Kolotyrkin and S. E. Kupriyanov, *Zh. Fiz. Khim. Akad. Nauk SSSR*, **37**, 2769 (1963) [English trans.: *Russ. J. Phys. Chem.*, **37**, 1498 (1963)].

126. S. E. Kupriyanov and A. A. Perov, *Izvest. Akad. Nauk SSSR, Ser. Fiz.*, **27**, 1102 (1963) [English trans.: *Bull. Acad. Sci. USSR, Phys. Ser*, 27, 1072 (1963)].

127. S. E. Kupriyanov and A. A. Perov, *Zh. Tekh. Fiz.*, **34**, 1317 (1964) [English trans.: *Sov. Phys.-Tech. Phys.*, **9**, 1018 (1965)].

128. S. E. Kupriyanov, *Zhur. Eksp. Teor. Fiz.*, **47**, 2001 (1964) [English trans.: *Sov. Phys.-JETP*, **20**, 1343 (1965)].

129. K. M. Refaey and W. A. Chupka, *J. Chem. Phys.*, **43**, 2544 (1965).

130. E. Gustafsson and E. Lindholm, *Arkiv Fysik*, **18**, 219 (1960).

131. V. V. Afrosimov, R. N. Il'in, and N. V. Fedorenko, *Zh. Eksp. Teor. Fiz.*, **34**, 1398 (1958) [English trans.: *Sov. Phys.-JETP*, **7**, 968 (1958)].

132. E. S. Solov'ev, R. N. Il'in, V. A. Oparin, and N. V. Fedorenko, *Zh. Eksp. Teor. Fiz.*, **42**, 659 (1962) [English trans.: *Sov. Phys.-JETP*, **15**, 459 (1962)].
133. V. F. Kozlov and S. A. Bondar, *Zh. Eksp. Teor. Fiz.*, **50**, 297 (1966) [English trans.: *Sov. Phys.-JETP*, **23**, 195 (1966)].
134. R. Browning, C. J. Latimer, and H. B. Gilbody, *J. Phys. B (Atom. Mol. Phys.)*, **2**, 534 (1969).
135. C. J. Latimer, R. Browning, and H. B. Gilbody, *J. Phys. B (Atom. Mol. Phys.)*, **2**, 1055 (1969).
136. E. Lindholm, *Arkiv Fysik*, **8**, 433 (1954).
137. C. F. Giese and W. B. Maier, II, *J. Chem. Phys.*, **39**, 197 (1963).
138. W. B. Maier, II, *J. Chem. Phys.*, **41**, 2174 (1964).
139. E. Lindholm, *Arkiv Fysik*, **8**, 257 (1954).
140. V. A. Gusev, G. N. Polyakova, and Ya. M. Fogel', *Zh. Eksp. Teor. Fiz.*, **55**, 2128 (1968) [English trans.: *Sov. Phys.-JETP*, **28**, 1126, (1969)].
141. R. N. Il'in, V. V. Afrosimov, and N. V. Fedorenko, *Zh. Eksp. Teor. Fiz.*, **36**, 41 (1959) [English trans.: *Sov. Phys.-JETP*, **9**, 29 (1959)].
142. W. B. Maier, II, LA-DC-8633, LASL (1967).
143. W. B. Maier, II, *J. Chem. Phys.*, **47**, 859 (1967).
144. H. Sjogren and E. Lindholm, *Arkiv Fysik*, **32**, 275 (1966).
145. V. M. Dukel'skii and E. Ya. Zandberg, *Dokl. Akad. Nauk SSSR*, **99**, 947 (1954) (English trans.: Redstone Arsenal Technical Library, Translation Nr 20-62).
146. V. M. Dukel'skii and N. V. Fedorenko, *Zh. Eksp. i Teor. Fis.*, **29**, 473 (1955) [English trans.: *Sov. Phys.-JETP*, **2**, 307 (1956)].
147. C. F. Barnett, M. Rankin, and J. A. Ray, *6th International Conference on Ionization Phenomena in Gases*, **1**, 63 (1963).
148. S. E. Kupriyanov, A. A. Perov, and N. N. Tunitskii, *Zh. Eksp. Teor. Fiz.*, **43**, 763 (1962) [English trans.: *Sov. Phys.-JETP*, **16**, 539 (1963)].
149. S. E. Kupriyanov, *Kinetika i Kataliz*, **3**, 13 (1962) [English trans.: *Kinetics and Catalysis*, **3**, 9 (1962)].
150. S. E. Kupriyanov and A. A. Perov, *Zh. Tekh. Fiz.*, **33**, 823 (1963) [English trans.: *Sov. Phys.-Tech. Phys.*, **8**, 618 (1964)].
151. E. Lindholm, *Z. Naturforsch.*, **9a**, 535 (1954).
152. H. Sjogren, AF 61(052)-762 (1966).
153. E. Lindholm, *Proc. Phys. Soc.*, **A66**, 1068 (1953).
154. H. Sjogren, *Arkiv Fysik*, **32**, 529 (1966).
155. V. A. Gusev, D. V. Pilipenko, and Y. M. Fogel, *Zh. Eksp. Teor. Fiz.*, **51**, 1007 (1966) [English trans.: *Sov. Phys.-JETP*, **24**, 671 (1967)].
156. W. B. Maier, II, *J. Chem. Phys.*, **42**, 1790 (1965).
157. Hitoshi Yamaoka, Pham Dong, and J. Durup, *J. Chem. Phys.*, **51**, 3465 (1969).
158. G. Sahlstrom and I. Szabo, U.S. Clearinghouse Fed. Sci. Tech. Inform. (1967) AD666027.
159. G. R. Hertel and W. S. Koski, *J. Am. Chem. Soc.*, **86**, 1683 (1964).
160. J. G. Collins and P. Kebarle, *J. Chem. Phys.*, **46**, 1082 (1967).
161. J. C. Abbe, *J. Chim. Phys.*, **65**, 472 (1968).
162. R. H. Schuler and F. A. Stuber, *J. Chem. Phys.*, **40**, 2035 (1964); **41**, 901 (1964).
163. S. Wexler, *J. Chem. Phys.*, **41**, 2781 (1964).
164. E. Lindholm, I. Szabo, and P. Wilmenius, *Arkiv Fysik*, **25**, 417 (1964).
165. C. E. Melton and P. S. Rudolph, *J. Chem. Phys.*, **30**, 847 (1959).
166. S. Wexler and D. C. Hess, *J. Chem. Phys.*, **38**, 2308 (1963).

167. S. E. Kupriyanov, *Zh. Tekh. Fiz.*, **36**, 2161 (1966) [English trans.: *Sov. Phys.-Tech. Phys.*, **11**, 1614 (1967)].
168. C. E. Melton and G. A. Ropp, *J. Chem. Phys.*, **29**, 400 (1958).
169. J. B. Homer, R. S. Lehrle, J. C. Robb, and D. W. Thomas, *Trans. Faraday Soc.*, **62**, 619 (1966).
170. H. von Koch, *Arkiv Fysik*, **28**, 529 (1965).
171. G. R. Hertel and W. S. Koski, *J. Amer. Chem. Soc.*, **87**, 1686 (1965).
172. H. Sjogren, *Arkiv Fysik*, **31**, 159 (1966).
173. E. R. Weiner, G. R. Hertel, and W. S. Koski, *J. Am. Chem. Soc.*, **86**, 788 (1964).
174. S. E. Kupriyanov and A. A. Perov, *Zhur. Fiz. Khim.*, **42**, 857 (1968) [English trans.: *Russ. J. Phys. Chem.*, **42**, 447 (1968)].
175. V. L. Tal'roze, *Izvest. Akad. Nauk SSSR Ser. Fiz.*, **24**, p. 1001 (1960) [English trans.: *Bull. Akad. Sci. (USSR) Phys. Ser.*, **24**, 1006 (1960)].
176. I. Szabo, *Arkiv Fysik*, **31**, 287 (1966).
177. J. C. Abbe and J. P. Adloff, *Compt. Rend.*, **258**, 3003 (1964).
178. J. C. Abbe and J. P. Adloff, *Phys. Letters*, **11**, 28 (1964).
179. P. Wilmenius and E. Lindholm, *Arkiv Fysik*, **21**, 97 (1962).
180. H. Sjogren, *Arkiv Fysik*, **29**, 565 (1965).
181. H. von Koch, *Arkiv Fysik*, **28**, 559 (1965).
182. H. von Koch and E. Lindholm, *Arkiv Fysik*, **19**, 123 (1961).
183. E. Pettersson and E. Lindholm, *Arkiv Fysik*, **24**, 49 (1962).
184. E. Pettersson, *Arkiv Fysik*, **25**, 181 (1964).
185. W. A. Chupka and E. Lindholm, *Arkiv Fysik*, **25**, 349 (1963).
186. G. R. Hertel and W. S. Koski, Adv. Chem. Ser. **72**, 15 (1968).
187. G. R. Hertel and W. S. Koski, *J. Am. Chem. Soc.*, **87**, 404 (1965).
188. G. W. McClure, *Phys. Rev.*, **166**, 22 (1968).
189. A. J. Akishin, *Usp. Phys. Nauk*, **66**, 331 (1958) [English trans.: *Sov. Phys. Usp.*, **66**, 113 (1958)].
190. P. J. Chantry and G. J. Schulz, *Phys. Rev. Letters*, **12**, 449 (1964).
191. A. Russek, *Phys. Rev.*, **120**, 5 (1960).
192. S. G. Rautian, *Sov. Phys.-Usp.*, **66**, No. 1, 245 (1958).
193. F. P. Ziemba, G. J. Lockwood, G. H. Morgan, and E. Everhart, *Phys. Rev.*, **118**, 1552 (1960).
194. S. E. Kupriyanov, *Kinetika i Kataliz*, **6**, 532 (1965) [English trans.: *Kinetics and Catalysis*, **6**, 460 (1965)].
195. W. A. Chupka, M E. Russell, and K. Rafaey, *J. Chem. Phys.*, **48**, 1518 (1968).
196. E. E. Salpeter, *Proc. Phys. Soc. (London)*, **A63**, 1295 (1950).
197. L. D. Landau and E. M. Lifshitz, *Quantum Mechanics*, Addison-Wesley, Reading, Mass. (1958), p. 419.
198. See the review by A. Burgess and I. C. Percival, in *Advances in Atomic and Molecular Physics*, **4**, edited by D. R. Bates and I. Estermann, Academic Press, New York (1968), p. 109.
199. C. J. MacCallum, private communication (1969).
200. V. I. Gerasimenko and Yu. D. Oksyuk, *Zh. Eksp. Teor. Fiz. (USSR)*, **48** 499 (1965) [English trans.: *Sov. Phys.-JETP*, **21**, 333 (1965)].
201. D. R. Bates, *Atomic and Molecular Processes*, edited by D. R. Bates, Academic Press, New York (1962), p. 549.
202. J. D. Craggs and H. W. S. Massey, in *Handbook der Physik*, edited by S. Flugge, Springer-Verlag, Berlin (1959), **37**, Pt. 1, p. 332.

203. T. Iijima, R. A. Bonham, and T. Ando, *J. Phys. Chem.,* **67,** 1472 (1963).
204. J. M. Peek, *Phys. Rev.,* **134,** A877 (1964).
205. D. R. Bates and A. R. Holt, *Proc. Phys. Soc. (London),* **A85,** 691 (1965).
206. K. J. Miller and M. Krauss, *J. Chem. Phys.,* **47,** 3754 (1967).
207. R. A. Bonham and T. Iijima, *J. Phys. Chem.,* **67,** 2266 (1963).
208. T. Iijima and R. A. Bonham, *J. Phys. Chem.,* **67,** 2769 (1963).
209. D. C. Cartwright and A. Kuppermann, *Phys. Rev.,* **163,** 86 (1967).
210. J. M. Peek, *Phys. Rev.,* **183,** 193 (1969).
211. J. M. Peek and T. A. Green, *Phys. Rev.,* **183,** 202 (1969).
212. T. A. Green, *Phys. Rev.,* **157,** 103 (1967).
213. T. A. Green and J. M. Peek, *Phys. Rev.,* **169,** 37 (1968).
214. E. N. Lassettre and E. A. Jones, *J. Chem. Phys.,* **40,** 1222 (1964).
215. M. Born and J. R. Oppenheimer, *Ann. Phys.,* **84,** 457 (1927).
216. E. H. Kerner, *Phys. Rev.,* **92,** 1441 (1953).
217. J. M. Peek, *Phys. Rev.,* **140,** A11 (1965).
218. T. A. Green and J. M. Peek, *Phys. Rev. Letters,* **21,** 1732 (1968).
219. T. A. Green and J. M. Peck, *Phys. Rev.,* **183,** 166 (1969).
220. J. M. Peek, *Phys. Rev.,* **139,** A1429 (1965).
221. E. U. Condon, *Phys. Rev.,* **32,** 858 (1928).
222. J. G. Winans and E. C. G. Stueckelberg, *Proc. Natl. Acad. Sci. U. S.,* **14,** 867 (1928).
223. A. S. Coolidge, H. M. James, and R. D. Present, *J. Chem. Phys.,* **4,** 193 (1936).
224. Yu, D. Oksyuk, *Opt. i Spektroskopiva,* **23,** 366 (1967) [English trans.: *Opt. Spectry. (USSR),* **23,** 197 (1967)].
225. G. H. Dunn, *Phys. Rev.,* **172,** 1 (1968).
226. Yu. D. Oksyuk, *Opt. i Spektroskopiva,* **23,** 213 (1967) [English trans.: *Opt. Spectry. (USSR),* **23,** 115 (1967)].
227. R. N. Zare, *J. Chem. Phys.,* **47,** 204 (1967).
228. J. H. Moore, Jr., and J. P. Doering, *J. Chem. Phys.,* **50,** 1487 (1969).
229. R. G. Alsmiller, Jr., Research Report ORNL-3232 (1962), unpublished.
230. E. Gerjuoy, Research Report WRL-NP-7170 (1955), unpublished.
231. J. M. Peek, *Phys. Rev.,* **154,** 52 (1967).
232. G. H. Dunn and B. Van Zyl, *Phys. Rev.,* **154,** 40 (1967).
233. D. F. Dance, M. F. A. Harrison, R. D. Rundel, and A. C. H. Smith, *Proc. Phys. Soc. (London),* **A92,** 577 (1967).
234. J. M. Peek, T. A. Green, and W. H. Weihofen, *Phys. Rev.,* **160,** 117 (1967).
235. J. Durup, presented at the 14th Annual Conference on Mass Spectroscopy and Allied Topics, Dallas, Tex. (May 22–27, 1966).
236. J. E. G. Farina, *Proc. Phys. Soc. (London),* **A90,** 323 (1967).
237. G. H. Dunn, *Phys. Rev. Letters,* **8,** 62 (1962).
238. G. H. Dunn and L. J. Kieffer, *Phys. Rev.,* **132,** 2109 (1963).
239. E. H. Kerner, *Phys. Rev.,* **92,** 1441 (1953).
240. T. A. Green, *Phys. Rev.,* **A1,** 1416 (1970).
241. M. H. Mittleman, *Phys. Rev.,* **137,** A1 (1965).
242. B. H. Bransden, *Advances in Atomic and Molecular Physics,* **1,** edited by D. R. Bates and I. Estermann, Academic Press, New York (1965), p. 85.
243. A. Russek, *Physica,* **48,** 165 (1970).
244. R. G. Alsmiller, Jr., Research Report ORNL-2766 (1959).
245. E. Bauer, *Phys. Rev.,* **84,** 315 (1951).

246. E. Bauer, *Phys. Rev.,* **85,** 277 (1952).
247. R. G. Breene, Jr., *Phys. Rev.,* **131,** 2560 (1963).
248. R. G. Breene, Jr., *J. Chem. Phys.,* **45,** 3876 (1966).
249. D. R. Sweetman, *Phys. Rev. Letters,* **3,** 425 (1959).
250. C. F. Barnett, *2nd Geneva Conf., on the Peaceful Uses of Atomic Energy,* **32,** 398 (1958).
251. R. V. Pyle, K. Berkner, S. Kaplan, and J. W. Stearns, *Bull. Amer. Phys. Soc.,* **9,** p. 426 (1964).
252. W. L. Fite, A. C. H. Smith, R. F. Stebbings, and J. A. Rutherford, *J. Geo. Res.,* **68,** 3225 (1963).
253. J. Guidini, *Compt. Rend.,* **252,** 2848 (1961).
254. J. Guidini, R. Belna, G. Brifford, and C. Manus, *Compt. Rend.,* **251,** 2496 (1960).
255. J. Guidini, *Atomic Collision Processes,* 751 (1964).
256. F. P. G. Valckx and Verveer, IV International Conference on the Physics of Electronics and Atomic Collisions, Quebec, Canada (1965). p. 333.
257. D. W. Vance and T. L. Bailey, IV International Conference on the Physics of Electronics and Atomic Collisions, Quebec, Canada (1965). p. 273.
258. M. Vogler, W. Seibt, and H. Ewald, V International Conference on the Physics of Electronics and Atomic Collisions, Leningrad, USSR (1967). p. 587.
259. T. Sinda, C. Manus, and J. Guidini, V International Conference on the Physics of Electronics and Atomic Collisions, Leningrad, USSR (1967). p. 597.
260. N. V. Fedorenko, R. N. Il'in, and E. S. Solov'ev, Proceedings of the Vth International Conference on Ionization Phenomena in Gases, edited by H. Maecker, North Holland Publ. Co., Amsterdam (1962), **II,** p. 1300.

INDEX